Bobcats Before Breakfast

JOHN KULISH *with* AINO KULISH

STACKPOLE
BOOKS
Guilford, Connecticut

Published by Stackpole Books
An imprint of The Rowman & Littlefield Publishing Group, Inc.
4501 Forbes Blvd., Ste. 200
Lanham, MD 20706
www.rowman.com

Distributed by NATIONAL BOOK NETWORK

Illustrations by Aino Kulish

British Library Cataloguing in Publication Information available

Library of Congress Cataloging-in-Publication Data available

ISBN 978-0-8117-3886-6 (paperback)
ISBN 978-0-8117-6894-8 (e-book)

♾ᵀᴹ The paper used in this publication meets the minimum requirements of
American National Standard for Information Sciences—Permanence of Paper for
Printed Library Materials, ANSI/NISO Z39.48-1992.

Contents

Matching wits with muskrats made more sense than struggling with fractions. The books I wanted to read were all in nature's library.

Any man who chooses to support a wife, two daughters, and assorted hound dogs by hunting and trapping in outer suburbia in the middle of the twentieth century needs all the help he can get.

A first-rate woods linguist can translate stories with complicated plots woven around several characters, written, not only in snow, but in frost, in leaves, and even in a few wisps of disturbed vegetation.

To a person who has learned, not only nature's mother tongue, but most of her dialects, a walk through the woods is like reading a great novel. Every few steps, new characters are introduced. The plot changes and develops; the tempo increases or subsides. Climax follows climax. One can hardly wait to turn the page.

Do you suppose a wood mouse says, "Some of my best friends are meadow mice, but-"? Do you suppose starlings envy the scarlet tanager his fancy "mod" jacket? Do you suppose sparrows attack crows because they are black? Animal behavior is logical; wild creatures don't know how to rationalize.

The first thing I learned was that gunpowder, two legs, and a half domesticated brain barely balanced a wildcat's four legs, eagle eyes, and radar ears. The bobcat's eyes do more than see, and its ears do more than hear; for its eyes and ears are a cat's "brains." They warn it of danger. They find it its food.

You can always tell a good hound. Like a good man, he earns his keep. He knows what he's supposed to do-and does a little more. Money can't separate the exceptional man-dog team any more than it can break up a good man-woman pair. The real thing is unbuyable, unbreakable, unbeatable. It ends only when one of them dies.

A dog carries its character, as well as its heart, in its eyes. The dark eyes, looking steadfastly into mine, were calm, honest, and bright with intelligence. But it was something else that glimmered in their depths that put a finger on my heart.

Kitty Guy's fame spread and, when he was seven, I refused a thousand dollars for him. "Would you sell your wife or your children?" I asked the would-be buyer. My wife summed it up cheer fully, "If Jiggs could cook, I'd be out of a job."

Climbing a steep mountain without snow is difficult. Climbing a steep mountain with snowshoes is worse. Climbing a steep moun tain in deep snow without snowshoes is a horrible experience.

As a woodsman, I frequently faced physical danger. When face to face, the fear I felt was usually sudden and short-lived. But that dark day on the mountain, fear of the unknown, fed by un certainty, made my very heart buckle.

A beaver pond lifts the horizons of the entire forest community. Within months, it becomes the civic center for inhabitants from miles around; a forest plaza, with varied restaurants; the most popu-lar meeting place for animals that enjoy good food.

The civil engineering curriculum of the apprentice beaver includes courses in Introductory Tree Identification, Advanced Selective Cutting, The Natural Philosophy of Water Pressures and Cur rents, Scientific Erosion Control, Tunnel- and Canal-building, Sluicing of Logs, Types of Mud and Clay, and Stream Flow Char acteristics.

As a man, I would like to be the kind of man an otter is as an animal. To him, life is an adventure. He is rarely foolish, but neither does caution dull his high spirits.

Watching until the otter family disappeared in a cove, still rollick ing, I knew none of their clan could ever develop ulcers or hyper-tension. Ever since that autumn day, whenever feelings of guilt about finishing a stone wall or cleaning the cellar assail me, I remember those happy acrobats and reach for my fishpole or my snowshoes.

In today's woods, a hunter on snowshoes is as out-of-date as a crossbow at a missiie site. As gracefully as I could, I hung up my traps and my guns. My privilege to hunt comes loaded with re sponsibility. I owe the game I pursue a place to live, food to find, and ample cover to hide from me. Even more, as a man, I owe them an even break.

Acknowledgments

The Harris Center for Conservation Education would like to thank the Kulish family for their support in seeing Aino's words and John's stories find the light of day again and Judith Schnell of Stackpole Books for her commitment to this book. Thank you, too, to Mark Reynolds for his heartfelt and evocative foreword.

John Kulish was legendary. His spirit of adventure, deep curiosity, and love for all things wild lives on in the work of the Harris Center, where we continue to walk in John's tracks, teaching people of all ages about the natural world.

Foreword

As the remarkable account of one man's extraordinary commitment to the natural world, *Bobcats Before Breakfast* stands alone. Unlike many nature writers who glean much of their knowledge from the classroom, John Kulish learned his lessons by pitting his wits, muscle, and endurance against a variety of furbearers and eking out an existence as a professional trapper, bobcat hunter, and hunting guide.

From boyhood, John was drawn to woods and wild places—a devotion that shaped his entire life. In the 1950s, when he realized that the modern world was encroaching too deeply into the wilderness, John hung up his guns and traps and stepped from his old realm into a future for which he felt completely unprepared. Assuming it was the best he could expect, he took a job as a janitor at Boston University's Sargent Center for Outdoor Education in Hancock, New Hampshire. The staff soon recognized his expertise and set him to work sharing his rare knowledge of the natural world with college students. It was a wonderful stroke of luck for a man who thought the way of life he loved had come to an end. "I'd go to bed whistling, and wake up whistling," he said of his new job.

After his stint at Sargent Camp, John became the naturalist at the Harris Center for Conservation Education, also in Hancock, where he taught in the local schools and led group hikes through the backcountry he knew so well.

The folks at the Harris Center and Sargent Camp weren't the only ones who understood the value of John's hard-won experience. His wife, Aino, a bright and free spirit born of the local Finnish community, had an artistic as well as a practical side. In addition to her success at stretching the family's meager income to keep food on the table for them and their two daughters, she was also a painter and writer. Her painting of John's beloved dog, Jiggs, hung over their mantelpiece. At the end of the workday, after supper, she would sit John down, pull out her notebook and pen, and say, "Start talking." *Bobcats Before Breakfast* is the result of those evenings.

I first met John Kulish on a snowshoe hike he led for the Harris Center in 1986. I was thirty-five and he was seventy-five. Though I'd heard of him—he was something of a legend in the local outdoor community—I hadn't yet read *Bobcats Before Breakfast*. First published in 1969, it was out of print by the mid-1980s; used copies were highly sought after, especially in this part of New Hampshire. I'm delighted that the Harris Center is now making John's story available to a new generation of readers.

Tall and lean, John was smart, tough, and insatiably curious about the outdoors and its inhabitants. Almost until the day he died (at age eighty-five), his long stride could keep pace and cover the distances through the woods, over the ledges, and around the swamps he knew so well. He never stopped to rest or eat on a day-long hike, though, for the sake of those who attended his Harris Center outings, he would halt at noon to boil water in a #10 tin can and brew up strong tea, giving everyone else a chance to sit down and eat whatever they brought for lunch. John never sat, but stood by the fire, mug of steaming tea in hand, regaling his listeners with his endless supply of stories and woods lore.

The hikers would question John about the things they saw—some, I'm sure, in hopes of stumping him. He always knew the answer. Not only could he identify every track, no matter how old or snowed in, he could tell what the animal had on its mind: if it were hunting, for example, or just passing quickly through an area where it felt vulnerable.

Once, at the end of a long day on snowshoes, John and I followed an old woods road down a mountain. Snowmobiles had packed the snow under our feet, and we walked along, chatting, when suddenly John looked down and said, "Those are bear tracks!" In the fading light, the tracks were barely discernible—just a few indistinct depressions crossing the road. How could he know those faint marks were left by a bear? "Look where they come from, and look where they go," John explained. There was a thick growth of whipstock and black raspberry canes on both sides of the road. "Nobody else would walk through there when he could go around. But a bear's a bulldozer; he doesn't care what he walks through."

The old woodsman had demonstrated once again that, to unravel one of Nature's puzzles, you have to look at the surrounding evidence as a detective does a crime scene, considering all the clues. Often what you don't see is as important as what you do.

The real amazement for me on that occasion was that John saw the obscured tracks at all. We were walking side by side at the end of a long, tiring day. He certainly didn't seem to be paying any closer attention than I was. But his years of earning a living interpreting animal sign had honed in him a keen and constant power of observation. At that time, there were fewer bear (and bobcats!) roaming the Monadnock Region than there are today, so the tracks were uncommon.

John also had an expert knowledge of ice, perfected from countless crossings of frozen lakes and streams where a fall through could have dire consequences. Once he had us cross a frozen beaver pond where the black ice seemed paper-thin, confident it would support us. I remained skeptical until we reached the opposite bank. Another time, we followed a coyote's tracks down a snow-covered brook. Suddenly, John warned: "Don't step there!" I poked the snow just in front of me and my hiking staff sunk into an icy slush. He had noticed the slush in the coyote track and knew it meant that the coyote had stepped near a soft spot in the ice.

But what really drew me to John's side, and consequently what is so engaging about this book, is the connection he gave me to the past. John was an enchanting link to a long American tradition and to bygone ways and times. Born in 1911, he grew up hard: His parents were immigrants from Lithuania, and his father was an abusive alcoholic. John's early life was marked by poverty and sporadic violence. That, coupled with his rugged and solitary lifestyle, made John someone who could be as challenging as he was charming. He took great pride in his hard-earned knowledge and could be gruff with other naturalists, especially those who hailed from academia, or anyone else who hadn't put in the miles but intimated that they had an expertise on par with his own. And woe to anyone who dared tell John he'd seen a mountain lion in the woods he knew so well!

Those with the right credentials, however, had John's respect and admiration. As a young boy, he sought the knowledge of reticent Arthur Leonard, a local foxhunter who had been a teenager during the Civil War. Evidently, Mr. Leonard was an uncommunicative Yankee, but young John must have impressed him enough to earn at least some of the old man's regard. Mr. Leonard passed his handmade otter board down to John, who speculated on the number of otter pelts that had been stretched and dried over its thin frame. John also sought the company of Arthur Eastman, a noted Maine guide, who, well into his eighties, still traversed the Allagash country, sleeping in a series of crude shelters while tending his trapline. One of the most memorable evenings of my life was passed in the company of John and his lifelong friend, Vic Starzynski, as they swapped tales of their boyhood haunting of a Gardner, Massachusetts, pool hall to eavesdrop on the local foxhunters who gathered there. John was hungry to learn from anyone who'd acquired knowledge the old way. *Bobcats Before Breakfast* is the account of a man continuing a tradition that stretches from the earliest days of our country.

Many today may have a problem with the fact that John spent years trapping and killing animals for profit (meager and exacting though that profit was). But John took his harvest from a landscape that was much more abundant than anything we experience today, and he hung up his gun and traps when he realized the environment no longer supported the richness he once knew.

Consider the story he tells in the last chapter of the book—the day when, as a boy, he and a companion roused thirty-eight grouse from a huge oak tree. Or, even into the 1950s, when he spied on thirty-two deer yarded up for the winter. And hear the sorrow in John's words for the passing of that world: "Today, when I strap on my snowshoes and head for the familiar ridges, it's as though I were going to a wake. . . . But now I am alone. My woods are dead, overhunted, barren of game, with only snowmobile tracks where deer and bobcats once walked." It certainly wasn't the lone hunter, such as himself, that wrought such a transformation.

I'll never meet anyone who knows the woods like John Kulish did, and this book is his recording of the lessons he'd mastered over many years and thousands of grueling miles. "I take pride in my ability to hunt and trap," he wrote. "I learned it long, hard, and well."

Mark Reynolds
Antrim, NH

Mark Reynolds is a retired writer and editor and proud Harris Center member. As a younger man, he followed John Kulish through his favorite haunts—from Vermont to Maine's Allagash country—delighted in John's lore and stories, and quickly learned that there is no such thing as "catch and release" when you land a brook trout in John's presence. The old bobcat hunter may have given up hunting and trapping, but he never lost his taste for fresh trout.

Preface

This is a book about the woods, the animals who live there, and one man's lifelong relationship with them. I have learned many things, some of which conflict with entrenched ideas; but what I have learned did not come out of some text book. The animals taught me.

I lived among them. I watched them; I listened to them; I pursued them; I laughed with them; I cried with them; I out-witted them one day, only to be eluded by them on the next, until I learned how they think. I started out to understand animals and ended up understanding myself.

JOHN KULISH

Hancock, New Hampshire

1. I Belonged in the Woods

The three Navy doctors stood in a half-circle facing me as I sat on an examining table in the Newport Naval Hospital. Three days of continuous probing, pinching, and peering into bowel, bladder, and brain was almost over. It was January, 1942. It seemed months instead of weeks since that dark Sunday when, en route home from a day of guiding deer hunters, I had stopped in a tiny New Hampshire village, to be told by a gasoline station attendant that the Japanese had attacked Pearl Harbor. Before the car was back on the highway, I knew what I had to do. For this woodsman, there would be no more traps to tend nor animals to hunt.

The senior physician, his shoulders and sleeves glittering with gold bars and insignia, stepped closer. A big hand tilted my head as he peered into each of my ears. It seemed to take him a long time. He motioned to the two junior officers.

"Take a look at this."

I could feel warm breathing on my neck as, examination lights glowing, they peered into each of the last two unexplored hatches. There was a long silence. Then the trio moved away from the table. Bits and pieces of agitated sentences reached me.

"Never saw anything—"

"What do you suppose—"

"We'd better—"

The commander approached the table, and swung around briskly to face me.

"Anything wrong with your ears?"

My throat went dry. "No, sir."

"How old are you?"

"Thirty in April."

"Ever had trouble hearing?"

"No, sir."

An orderly quickly carried out the order to bring syringes, soap and water, and a pan. While a junior officer held the kidney-shaped dish under my ear, the commander himself fired the first salvo. Rushing noises filled my head. Another charge followed, then another. Again and again, the doctors squirted water into my resisting ears. Suddenly, with a loud gurgling, the bastion crumbled. Water splashed into the pan. From the corner of my eye, I saw the commander's face as he barked out, "Where the Hell have you been?"

There, in the enameled pan, floated the story of my life: pine needles, spruce and hemlock needles, twigs, leaves, and bark. It had taken twenty years as a woodsman to collect that panful of forest debris.

Most people rightfully think the professional woodsman, the man who hunts, traps, and guides for a living, moved west from New Hampshire with Horace Greeley. But, even though born a hundred years too late, while still a boy, I knew I belonged in the woods. By the time I was twelve, nearby rivers, brooks, bogs, and marshes were my home away from home. Matching wits with muskrats made more sense than struggling with fractions. From March to November, my boots and pants were caked with mud, and the rich redolence of stagnant bog water and animal musk surrounded me. I was the delight of every dog in town.

During my years in high school, from October until May, I went to school for but four days each week. The books I wanted to read were all in nature's library. It took several years to figure out her cataloging system, but once learned, I began to search out the rare volumes on her shelves and to learn to read her kind of printer's ink. Before long, I realized that if I persevered, sooner or later, a first edition might open before my wondering eyes. During four decades I made quite a collection of first editions. They cover all the native animals, many of the birds, the trees, and the plants, but, like any collector, I have favorites. Mine is the river otter, a volume so precious I have yet to share all its pages with anyone.

The old Yankee adage, "Give a boy a dog and a gun and you have the makings of a lazy man," has its Lithuanian counterpart. To my immigrant parents, the fulfillment of the American dream for their sons was a steady job in a furniture factory. Those were still the days when a man gladly worked sixty hours, with no coffee breaks, for ten dollars a week. For me, my parents' dream was a nightmare. During the rush periods of several summers I did work twelve to fourteen hours daily at a machine, and hated it. Insatiable, heartless, cast iron and steel, it threatened to hamstring my personality. How can anyone appreciate the sun's daily vault into the heavens, or calculate the progress of its steady march from a factory window?

The outdoors, the air, the wind, the clouds, the rain, the sleet, the snow, and the cold all mattered to me. I wanted to feel them all day, every day. I wanted to smell fir trees, bog water, and wild flowers. I wanted to watch a fox catch grasshoppers, see a doe nurse her fawn, spy on a hunting mink, and keep vigil on playing otter. I wanted to hear hawks whistle, partridges drum, brooks chuckle over stones washed down by some ancient glacier, and ice booming in deepening February cold. I wanted to follow the baying of a good hound. I wanted to be part of the winter, its clarity, its isolation, its rigorous demands. Each night, as I fell asleep, I wanted to be impatient for the morning, instead of dreading it as a quarry slave.

Purpose grew with muscles and bones. A boy's hope became a man's ambition: to learn all I could about the natural world I lived in and about the furred and feathered creatures that inhabited it with me. What makes a deer a deer, a mink a mink, a bobcat a bobcat, an otter an otter?

What I wanted to know would take my lifetime. It needed to be learned firsthand, in personal encounters with the animals and birds themselves. Just as the "proper study of mankind is man," so the only way to study wild animals is in their own element, without their being aware of your presence. To learn for myself what no books could teach me, my hands must hold and examine not one, but many, of the same species. The truth must be discovered by my heart as well as by my head. No endowment or foundation could be expected to support my practical yet unconventional research, but then, neither would I have to report to a doubting board of directors.

But how would this be possible? I must have food and shelter. Caves are damp, and rock tripe and birch polypores do not appeal to me. I wanted a home, a wife, and children. Still, I wanted to get my education from the forest, the marshes, the rivers, and from the wild creatures that flew, swam, and hunted there. Obviously, I could earn my "woods degree" only if the animals would pay my tuition.

My primary and secondary schooling in woodcraft, canoeing, hunting, fishing, and trapping progressed concurrently with the duller, conventional courses taught inside brick buildings. But my "Bachelor of Animals" took much longer than the usual four years. By the time nature awarded me my "master's," familiar mountains had grown taller and steeper.

Having worked for her before on a part-time basis, I knew exactly what I could expect from nature as my future employer. Demanding, immutable, unappeasing, yes; but also beautiful, pure, honest, exciting, and, best of all, completely trustworthy. Certain conditions always equal certain results. That is nature's greatest strength. There is no taskmaster or teacher like her. Once learned, her lessons are never forgotten. She demands that you observe, listen, think, compare, work, and remember. In turn, your hands, your feet, your brain, your heart, will bear blisters. Nature makes no concessions. She neither arbitrates nor bargains. A tough boss, she makes all the rules and none of the adjustments. For me, the one about a minimum wage proved the most rigorous.

For thirty years, we lived close to the bone, so close the marrow was often in jeopardy. We proved conclusively, and sometimes reluctantly, the economic theory that it is easier to live within a smaller income than within a larger one. In a vast, old house, heated with saw-it-yourself beaver-pond wood, one rarely develops a taste for satin sheets. But self-imposed economic discipline, invoked for a principle, usually pays compound interest as well as unexpected bonuses. The most obvious being a firm, flat belly. Dedication is low in calories. So is wild game.

We lived off the land. The big, dirt-floored cellar of our Victorian mausoleum smelled of stored potatoes, beets, carrots, and cabbage. Braced wooden shelves held row upon row of wild blueberries, strawberries, and blackberries, as well as vegetables and pickles. A freezer bulged with venison, hare, grouse, wild ducks, and horned pout. After her first week in school, when asked how she liked it, our older daughter, Johanna, replied, "Oh, I like Miss Wallace fine, but I don't like hot lunches. One whole week and we haven't had deer meat once!" Thanksgivings, as many as fourteen immigrant and first-generation Americans sat down to individual stuffed partridge, garnished with wild bog cranberries. Laughingly, my wife commented on how fitting it was that three hundred plus years after the initial feast, a first-generation American, married to an immigrant, provided for his family like a Pilgrim.

The months from October to May were the most important to me. During the frozen days of the longest shadows, the animals I pursued shared a common denominator with me—that of survival. The struggle to stay alive was as pitiless for them as it was for me. October meant ruffed grouse, woodcock, and ducks for the freezer, as well as scouting for fur. Each lake, pond, river, and brook in my district had to be surveyed. Where would the mink and the otter be? On which watershed should I begin? Where could I use a canoe? Where would I have to walk? Like a capable general, I drew my battle lines, while reconnoitering from dawn to dusk. Strategies were studied and restudied before D-Day dawned on November first.

During the next four weeks, an imminent freeze-up hung over my head like the sword of Damocles. Each morning, hours before daylight, my pack, loaded with traps weighing as much as seven pounds apiece, was hoisted onto my back. A shotgun in hand, I strode through the woods, "stringing steel," until long after dark. I never sat down to rest. There wasn't time. I stopped only to end the jetlike burst for freedom of a partridge or a duck, sometimes to watch a startled deer. I learned to rest while standing, but I was usually too wet with sweat to stop long.

My pack rarely lightened, for whenever I removed a trap, sooner or later, a twenty-pound otter took its place. One or two hours after dark, the soaked pack basket would be swung off my shoulders and into the back of my little truck. At home several mink and otter waited to be skun and stretched. The nights were too short. So were the days. There were never enough hours to do all that needed to be done: the cumbrous loads to be carried, the rugged miles to be hiked, the swamps to be waded, the rivers and lakes to be paddled. They were all part of a day's work, work that was fulfilling, exciting, and right for me.

My purpose was not just to catch a water animal and to stretch its pelt onto a board prior to selling it to a fur dealer. My aim was to learn as much as I could about every wild creature. So its domain became my domain. Did it have a permanent home address? Or was it a nomad? What did it do in the spring, the summer, the fall, and the winter? Where did it hunt? What did it eat? When did it kill? Did it fight with its relatives? With other creatures? When did it travel? Did it communicate with its peers? If so, how? Over the years, I learned to think like the animals. Their weaknesses became my strengths; their strengths, my weaknesses.

My motives, my *modus operandi,* my complete emotional, intellectual, and physical involvement alienated me from some part-time hunters and trappers, from braggarts, and from my mother-in-law. Game wardens without in-depth knowledge of animal habits, never could quite understand how it was possible for me to check an otter trap from a canoe thirty yards away. As a "woods scientist" and a "pay-as-you-learn naturalist," publicity was anathema to me. I avoided it as I would U.S. Route No. 1. My goal remained a simple one: to earn my living and my learning from the woods. It did not include becoming a millionaire.

In the woods, I was always hungry because eating took too much time. Besides, frozen apples, oranges, or sandwiches aren't worth the dental agony involved. But I seldom got discouraged—only angry at myself whenever I made a bad decision. How could I get discouraged? I belonged in the woods; they belonged to me. The mountains and the valleys for miles around were as familiar and as dear as were the rooms in my home. The animals were all related to ones I had already met. To me, there was no mystery in where a cat, a mink, or an otter would go. If one of them eluded me today, it would only delay our eventual face-to-face encounter tomorrow, or the day after.

While in the woods, I frequently discussed my problems with myself. Looking up at a ledge, I'd say out loud so that the wind could bear witness, "Well, Tom, you may have pulled the wool over my eyes today, but tomorrow is another day. I know you hate my guts, and I don't blame you. But I'm giving you fair warning. I know where you'll be, and I'm coming after you." Before daylight the next morning I was back where we had left off.

I had no more illusions about my chosen field than any competent captain of industry has about his. Like him, I acquired the wisdom to roll with the punches, and to plan a counterattack while still on my hands and knees. For mine was a battle of wits and of legs. Whenever the balance of power dipped in favor of my brains, it soon jerked back in favor of my quarry's sheer physical power. No man can outrun a cat or a deer or outswim an otter. Even a special man with special knowledge has his hands full trying to outwit an animal. Four legs are still better than two, especially in mountain climbing; and any wild animal's senses are far keener than ours. They still use a built-in system we dropped when we jumped, or fell, out of the trees before the Garden of Eden.

No matter how tired I was, nor how much I hurt, I never gave up. This was my field. I understood it. I understood myself. I believed in myself because I believed in what I was doing. I knew I could prevail, even though sometimes, it might take longer than others. What if things went wrong today? I had tomorrow. So did the animals.

2. Tools of a Woodsman's Trade

A professional woodsman's basic tools are few and simple: a short-barreled shotgun, a canoe, a compass, a jackknife that fits into the palm of his hand, snowshoes, all-wool underwear (in case he has to sleep in a snowdrift), a pack basket that bears the curve of his back, and boots with long laces.

To a professional hunter and trapper his boots are what a scalpel is to a surgeon or a saw is to a carpenter. Their make and condition tell you all you need to know about a hunter plying his trade. His boots are an outdoorsman's trademark. Just as any old saw is not found in the village artisan's tool box, so just any old kind of boots are not found drying on the hearthstone of a true woodsman.

I earned my living with my mind, my aim, and my feet. Day after day, month after month, year after year, I walked, climbed, ran, and jumped from daylight until dark. Whether it rained, snowed, sleeted, or shone, I waded brooks, crossed rivers, leapfrogged through swamps, wallowed in snow, slipped over ice, and sloshed through slush. A day that began with my snowshoes skimming over frozen crust could end up ankle-deep in water and thigh-deep in wet snow. To cover ground efficiently, a professional woodsman's feet must be as warm and limber as a centipede's.

Because we rarely have two consecutive days of identical weather, New England climate taxes a woodsman's footgear even more than it does his disposition. There have been days in the woods when I could have used everything from hip boots to sneakers.

Even when on the original producer, leather is not completely waterproof. Furthermore, leather has an appetite and finds wet snow irresistible. All-rubber boots were designed for standing in sewers, while insulated boots must have been fashioned with the Abominable Snowman in mind. There is only a single type of boot with which one is able to snowshoe thousands of miles, ford countless streams, wade through bog after bog, and forget he has feet.

Invented fifty years ago by a Maine manufacturer-outdoorsman, the rubber-bottomed, leather-topped hunting pac is still the best all-round boot for any woodsman who walks from October to May. (If you plan to spend your outdoor hours zooming over snowdrifts in a snowmobile, it won't be your feet that will trouble you.) Laced up from the instep, "rubber bottoms, leather tops" give solid support to a shock-absorbing ankle and lower leg. Most important to professional and amateur alike, they are amphibious.

Several times during a single day, I had to choose between crossing a knee-deep stream or hiking many miles cross-country to a better ford. I never hesitated. One, two, three quick tiptoe strides took me across, where I flopped down on my stomach and arched my legs skyward as though in a swan dive. Extra-long rawhide laces, wrapped snugly several turns around the top of my twelve-inch-high boots, prevented water from seeping in. For the rest of the day, my knees might be a little damp, but I didn't walk on my knees.

Those same long rawhide laces have many times spared me a night in a snowdrift. For some unfathomable reason, the leather straps that join the sandals to a snowshoe usually break just before dark, a dozen miles from a plowed road. A quick, temporary repair can be made with a jackknife and a piece of boot lace. After punching a hole through the end of each of the two broken straps, hold them together, worm the stolen strip of rawhide through the holes, and tie a snug square knot. I am sure every Cree Indian brave kept his squaw busy fall and winter chewing caribou leather thongs.

When worn with a removable felt innersole, "rubber bottoms, leather tops" are warm enough to withstand the worst a New England winter can offer. Each night the felts should be dried out. This may present a problem: wives are apt to react to fetid felts on a mantelpiece.

Light, flexible, waterproof, tough, versatile—and to top it off, these boots are economical. This factor alone endears them to the Yankee character. Even though I wore out two pair each year, because the worn-out bottoms could be stripped off and the leather bottoms returned for rebottoming as many as three times, my always ailing, pay-as-you-learn budget survived.

A powerful feeling stirs whenever I think about the Cree Indian who first designed the type of snowshoe I wear. He and I could have been blood brothers. Like him, on long, narrow snowshoes, for thousands of miles, I followed animal tracks, tracks that told me where the animal went, what it did, where it hunted, what it ate, where it defecated, how long it slept, when it mated, how long the marriage lasted, how it avoided its enemies; sometimes how it died.

One cannot learn animal habits from the seat of a speeding snowmobile, because the gasoline engine is still not the key that unlocks the secrets of the wild. Snowshoes and grit are.

There are few equally good ways to tune your physical fiddle as on the strings of snowshoes, especially when more miles than hours lie ahead. Under normal conditions, it is exercise which gladdens a woodsman's heart. On two strips of magic carpet, he can sprint over swampy terrain, impassable on foot or by canoe, once the snow and ice are gone. With ideal conditions underfoot and overhead, snowshoeing can be as lively as downhill skiing, and as rigorous as doing it cross-country.

Snowshoes come in a variety of styles, crafted to fit the local conditions. Almost every American Indian tribe uses its own individual design. For the New Hampshire open forest, undergrowth, and mountainsides, my preference is for a pair that is five feet long, ten inches wide, with definitely turned-up toes. Oftentimes, they substitute as short skis.

For four months a year I practically lived on snowshoes. Up mountains, down steep grades, across ledges, over frozen bogs, in and out of blowdowns, I averaged at least a dozen miles a day, every day, in all kinds of snow, and in all kinds of weather. Sometimes, miles away from the nearest plowed road, a sudden midday thaw would catch me unawares while still following a cat track. Past experience would turn my snowshoes back toward civilization. After the first four or five miles in that kind of going, my stockings felt wet. That night, when my boots came off, heavy woolen stockings stuck to my feet where red splotches had soaked through and dried. The five-hour struggle with spring snow had ground skin from winter-developed calluses on bony feet. If only I could have let the bobcats sleep, unmolested, for a few days, while my boots sat behind the kitchen stove.

The snowshoes got a rest because I had two pair, used on alternate days. But there was no rest for me. Next day, with or without the partial comfort of bandages, I was back in the woods. It was my livelihood. Even working only half-time, my snowshoes didn't always finish out the season with me. Sometimes, I had to borrow a pair. My own hung on a hook in tatters.

Like my snowshoes, my compass was my daily companion. It was often my lifeline. Even in familiar country, with the sun overhead to point the way, I reached up to feel the instrument's round reassurance in my shirt pocket. There, it shared honors with a folded topographic map of the immediate area. In strange country, a map and a compass go together. A compass without a map is like a man without a woman: he can go in a straight line, but rarely in the right direction.

Many sportsmen question my unequivocal reliance on a compass. "How come, John? We never use one."

With a straight face I answer, "I wouldn't need one either, if I never left a road."

Even for an experienced woodsman, being caught in a blizzard in deep woods can be an uncanny experience. Trees, sky, landmarks disappear. Only you and the storm remain. Unexpected snowshoe tracks become visible ahead. "Must be another hunter out." It takes a few minutes to realize there probably isn't another soul outdoors in this tempest in the state of New Hampshire. "Must be a game warden . . . but how come? Wardens seldom get out of their cars . . . Funny, he has snowshoes like mine, long and skinny . . . Look, mine fit right into the tracks!" Because you are worn out, it takes a few moments for the truth to penetrate: those are your own tracks. You've been moving in circles.

Your belly is empty. It's getting dark. Your shotgun weighs twice what it did at daybreak. You've already wasted a half-hour. Out comes your compass. You are through fooling around.

To a professional hunter, a shotgun is a tool that must be carried for thousands of miles in all kinds of weather, over all types of terrain. There are no hired gun bearers on snowshoe safaris. Therefore, the weapon must be light, with the shortest legal barrel obtainable. It must be of top quality. Even the best guns break under constant, hard usage. A gun's outward appearance reveals its role in a hunter's life. Toward the end of the winter, a professional woodsman's gun looks like a war souvenir, a relic from the Battle of Guadalcanal. The barrel, often without blueing, is dented and scratched from wild trips down ice-covered ledges. Chipped and battered, the stock and forearm have taken the brunt while its owner bumped into boulders and grabbed onto trees while hurtling down mountainsides. Blistered, cracked, the varnish is gone, a victim of too many ice, sleet, and snow storms. But the action, the working parts, tell the real story. Spotless, they are always oiled and in perfect working order.

To any man who matches his wits with the woods for a living, his jackknife and his trail axe are among his most precious possessions. His wife soon learns they are sacred ground. But there is something in the feminine psyche that cannot understand a man's feelings about razor-sharp, honed edges, and that even the toughest woodsman is sentimental about his knife and axe. My wife discovered the power of this emotion when my little jackknife, never more than a hand's reach away for thirty-five years, inadvertently accompanied a pair of trousers to the cleaner's. The owner of the establishment still talks about being awakened at four o'clock in the morning by the telephone to hear a woman's voice ask if he was interested in helping to prevent a divorce.

That little blade has skun the gamut from a two-ounce weasel to a 400-pound black bear.

Even more important than the tools a woodsman can buy, are the inborn aptitudes which the man alone can recognize and develop. The most vital of these is a magnetic attraction to the woods so powerful it borders on becoming an obsession. It makes him oblivious to barbs from friends and relatives, as well as to pleas from Conformity, who, tomahawk raised, pants on his heels, eager to add another scalp to her already trophy-studded belt. It even makes him oblivious to the weather. In New Hampshire, this alone is a test of character. Cold, ice, gales, sleet, snow, rain, all are secondary. The passion for knowing, no matter what the price, overpowers all else.

Size means nothing—except for the man's heart. That should be extra big, a twelve-cylinder engine fired by high-octane grit. I have yet to see a big, burly man who had endurance that kept him going hours after he should have collapsed. The tough ones were spare men. Some looked slight, one or two even frail. But their muscles were made of juniper, their lungs of nylon. They picked a pace and kept it: rough, easy, dry, wet, slippery, rocky, icy, level, steep. As such men move, one cannot hear their breathing. They never pant, nor do they sit down to rest. Their eyes are always moving. They see everything worth seeing and something more besides. Following a true woodsman in his own element makes it easy to see how an Indian moved and why. He sets down his worn hunting pacs like a starving fox sneaking up on a fat field mouse.

Grit is an invaluable quality for any man, but particularly for one who grapples with nature. It is grit that keeps him going when his feet are bloody. It is grit that refuses to accept defeat long after common sense has surrendered. It is grit that plunges him voluntarily into a backwoods river in near-zero temperatures.

With each passing hour, ice threatened to cut my economic lifeline. During the night, yesterday's storm clouds, hiking across the mountains, had grown weary of their load and dumped five inches of snow onto the frozen ground. Already, week-old ice covered the stream's edges. Now, in sudden, bitter cold, I could almost hear the main current freezing.

Yesterday, a trap's six-foot-long chain had been fastened to a stout piece of driftwood near shore. Both the driftwood log and trap were gone. Out in the middle, the log bobbed with the current. Obviously, it was moored to something. Previous experience sent me back into the woods where I cut down a tall, slender, dead spruce, and trimmed off the branches. With the pole I tried to reach out, to catch hold, and to haul in the driftwood. The log did not budge.

How easily a canoe would resolve the problem! But my canoe lay hidden in underbrush on another watershed twenty miles away.

My pack was unshouldered, my boots unlaced, my clothes stripped off. After a last, locating look, I sucked my lungs full of air and dove into liquid ice. With my eyes open, I spurted at an angle down toward the bottom.

I saw the chain first, then a drowned otter. It had wound the chain around and around an exposed tree root. With fingers that didn't seem to belong to me, I unwound the chain while my lungs begged for air. I lunged to the surface and, dragging the trap, swam back to shore. The snow felt strangely warm to my feet. Like a dog, I shook myself. Blessing my 100 percent all-wool underwear, I pulled on my clothes, relaced my boots, put the otter and trap into the pack, picked up my shotgun, and headed downriver.

At the time, I thought nothing of it; it was part of a day's work. But suppose someone had wandered into that isolated area that near-zero morning. What would he have thought to suddenly see a man standing, stark naked, in five inches of snow?

In the woods, playing a melodious second fiddle to grit, is a sense of humor. Without it, no man can work long for nature, for the odds are too one-sided. To a man miles away from civilization, a minor incident can become a complete catastrophe. A broken firing pin, a sudden late November squall that catches him in the middle of a mile-wide lake in an overloaded canoe, a dead bobcat that falls into the crotch of a towering tree on its downward arc, necessitating a fifteen-mile round-trip hike to borrow a rusty axe from a suspicious farmer, followed by a day of hand chopping to down an ash tree four feet in diameter. How could he do it if he couldn't laugh at his own helplessness? To survive, a woodsman must have the ability to lay, spent, on a snowdrift, his body soaked with sweat, his legs like boiled spaghetti, and chuckle, with only his hounds and the stars to hear him, over a day-long comedy of errors that netted him only further insight and a belated belly laugh.

In addition to the tools he can buy, and the innate peculiarities he must possess, there is a third factor, usually a question of pure luck, that is an absolute requisite for a woodsman's success. It is the acquisition of a resourceful wife. It goes without saying she must be a financial wizard. She also needs to have the diligence of an Iroquois squaw, the strength of a Percheron, the courage of a crusader, the disposition of an angel, and the thick hide of an elephant.

She needs to be a woman who can laugh when an acquaintance persists, "But what does your husband *do* for a living?" A woman who, when the larder is empty of staples, climbs into a shed loft, grabs a bushel basket brimming with porcupine heads, some ripening since October, and trundles them to the home of a selectman-judge, to collect the fifty cents per head bounty. A woman who, while chairmanning a PTA policy committee meeting, can ignore the sudden green faces of her colleagues when the piercing pungency of mink musk seeps up from the cellar where her trapper husband is skinning out next week's groceries. A woman who can laugh when a cauldron, filled with traps steeped in a rank, inky mixture of hemlock twigs and cedar shavings, bubbles over onto the burnished surface of her woodstove and thence onto a kitchen floor.

A woodsman's wife needs to know how to handle the other end of a crosscut saw, and can, in a pinch, buck up an evening's supply of wood, dangling earrings flying in rhythm with the saw's motion. She needs to keep dinner warm for hours, for she never knows when her husband will get home. She needs to be a

woman who, when she runs her husband's most cherished compass through the wringer of a washing machine, has the good
sense to dull the final agony by feeding him wild duck with wine
sauce, followed by hot apple pie, before confessing to her crime.

I can't take full credit for my good fortune because, when the
time came to choose my thirteenth rib, a benevolent Creator
gave a diffident backwoodsman a shove in the right direction.
Tall and straight, she looked full into my face and grinned.

"Hello, how's the duck hunting?"

Those five words hauled down my loner's flag forever.

In the intervening decades, full of learning, living, and loving,
I have never once doubted the wisdom of either choice, my work
or my woman, even while punching another hole into my belt
and wondering where the next mortgage payment was coming
from.

Fortunately, God looks after, not only fools and little children, but his mavericks as well. Any man who chooses to support
a wife, two daughters, and assorted hound dogs by hunting and
trapping in outer suburbia in the middle of the twentieth century,
needs all the help he can get.

3. Animal ABC's: Stories Wild Creatures Write

Every night, dramas are written in the woods. Tragedy, comedy, irony, mystery are all set down there waiting to be read. Even though most of the plays are nocturnal, we do not have to see them to know they happened. The wild animals themselves tell us. The theme of each of these stories is survival: I hunted; I found; I ate; I live. I was hunted; I was found; I was eaten; I died.

31

Every animal that walks, every bird that flies, knows how to write. They all leave notes. Some even leave letters. Wild creatures do not write as we do, for some of them write with their teeth, some with their claws, some with their wingtips, and some even with their tails. All of them write with their tracks and with their scat.

Big men sometimes walk with little feet, while small men often walk with big ones. But animal tracks never lie. Big animals have big feet; small animals have small ones. Long-legged animals take long strides; short-legged creatures take shorter ones.

Whenever I read an animal's tracks, I know immediately what it was, whether it was a big or a little creature, whether or not it was in a hurry, how long ago it had been there, where it had come from, and where it was going. Learning how to read animal writing is no different from learning how to read human text. When one first learns how to read, he begins with simple stories told in simple words. In reading animal and bird script, one should begin by learning how to decipher simple words written in an inch or two of soft snow: "I am a fox." As the subtleties of nature's semantics are mastered, more complicated scripts can be understood. The same writing will say, "Last night, I was here at midnight. The wind was blowing from the northwest. I found a field mouse under the snow. It tried to run away, but I caught it. I ate it. It was delicious, my favorite kind. I am still hungry, so I'm going to skip up to the old Thatcher place. Maybe the wind blew some apples down from a tree."

A first-rate woods linguist can translate stories with complicated plots woven around several characters, written, not only in snow, but in frost, in leaves, and even in a few wisps of disturbed vegetation. Translating a foreign language is a painstakingly slow, illogical process for most students. It's the darned grammar and the back-end-to phraseology that throws us off. It takes many years to become an expert at deciphering woods hieroglyphics. But then, does one expect to read the Aeneid during his first lesson in Latin?

To a woodsman, all picture writing is called sign. Out of the woods, there are many kinds of signs: road signs, advertising signs, warning signs, weather signs. In the woods, there is animal and bird sign. It is this the professional woodsman or a naturalist reads and interprets. Game sign is made by all birds and animals that are hunted. Fur sign covers all those wild creatures taken by trapping.

Whether it be game or fur, to me, animal sign is tracks: tracks in snow, in mud, in sand. It is scratches: scratches in sand, on logs, or on trees. It is bits of hair or of feathers. It is pieces of bones. It is drops of blood. It is a disturbance among leaves. It is sticks and tips of bushes stripped of bark. It is bubbles under ice. Above all, it is droppings; for scat is the universal biographer.

What kind of creature passed by? When? Where had it been? Where was it going? In what bistro had it dined? Had it eaten a three-course steak dinner, a blue plate special, or had it filled up on pretzels at the bar? Was it alone, or did it have a companion? And often most important of all, how big was it? To a consummate interpreter, scat reveals all.

A "woods scientist" always examines scat. To an experienced "bush pathologist," its size, its shape, and particularly its location are definitive. Most wild animals are fastidious in their personal habits. Each has its individual privy. Foxes, cats, coons, deer, mink, beaver, and otter leave their scat in certain places. The sloppiest creature is the porcupine. Its digestive juices must boil and bubble constantly, for it leaves a continuous fecal trail.

To examine an animal's scat is to read its diary. Fresh raccoon droppings found on the shores of a river or a lake in mid-October reveals that El Bandito has been living more like King Farouk. Yesterday, he had been up in the hills for acorn hors d'oeuvres. Then, en route to the fish course of frogs' legs and freshwater clams, he had stopped at a pioneer cellar hole for an apple and a pawful of grapes. Then he got his courses confused. But how could he resist crude honey? He had had a ringed eye on that hive of short-faced wasps since July. Only a fool would try to dig up ground hornets then. (They have no sense of humor about anyone who interferes with their papermaking.) He waited until a killing frost paralyzed his favorite dessert. Raccoons eat hornets for the same reason we eat corn. The insects are only partly digested.

In late fall and early winter, the scat of the black bear tells us that theirs is also a crude digestive machinery. Next to honey, black bears fancy apples. Because they sense the approach of hibernation, they eat the fruit compulsively. In order to keep up with the intake, intestinal conveyor belts run at high speeds. Little damage, beyond being halved and quartered, is done to apples.

Otter scat fascinates me. I have hiked many miles out of my way to read a river dog's potential travelogue. Because they travel constantly, mostly by water, sometimes by land, otter live exciting, adventuresome lives. Because they never have to fight for food, they play together every day. No otter is ever too old, too dignified, or too fat to play tag or touch football. Of all wild animals, they have the best minds and the warmest hearts.

If you are an expert at reading sign, you would know, as I did, that two days ago, a family of four stopped to fish and to visit. There was a large, fully mature dog, his wife, and their teen-age twins. Father had come ashore first. After a thorough investigation, he had scratched up leaves, pine needles, and humus into a pile. He had deposited his scat on top of the little mound. Then he signaled mother and the children that it was safe for them to disembark.

The power of suggestion works for them as it does for us. The size of each of the four mounds of scat identified the family members for me as surely as if I had been there. They had all been eating dace and little frogs. But it was the few pink slivers of shellfish that climaxed the story.

Two miles from where I stood, lay a small pond. It crawled with crayfish, but it belonged to a different watershed. From here, it could be reached only by an overland hike across hilly, rough terrain. Obviously, this household had made the difficult portage to the tiny pond before continuing their scheduled journey. Doesn't any trip go better when one has been fortified with fresh lobster?

Some of the most brain-stirring, heart-quickening dramas I have witnessed, were written into early winter's clear, "black" ice. And how can any animal write on ice, you ask? They don't write on the surface, but rather on the underside. It's like reading through a pane of glass.

All fur-bearing animals—mink, muskrat, beaver, and otter—can breathe under ice. This is a proved scientific fact. But how? Based on a lifetime of observation and investigation, I cannot agree with the popular belief that air is locked under ice. If this were so, the air would show in the form of bubbles. When ice is thick and covered with snow, bubbles can't be seen anyway, but no trapped bubbles show under the first clear ice of early winter. The only way to get air under ice is to put it there. To prove my theory, pick up a stone the size of a baseball and toss it up into the air over clear ice a quarter- to a half-inch thick. As it comes down, it will break through, sucking in air. Now watch small, white bubbles form. You put them there. Natural bubbles are made by a water animal's breathing or by bottom vegetation decaying to produce methane gas. Unlike human beings, water animals have been adapted to live part of their lives underneath ice. While there, they continually exhale dibs and dabs of their air supply, always holding a good portion in reserve. Contrary to accepted theories, I believe any one of them can remain under ice indefinitely, needing only to come out for food. My theory is that, whenever a water animal must replenish its air supply, it pushes its nose close to the ice and expels all its remaining air in the form of carbon dioxide. This instantly forms into a white bubble on the undersurface of the ice. Putting its nose back into the newly born bubble, the animal breathes in oxygen. I think a chemical change is somehow wrought by the ice itself and these original "frogmen" can swim merrily on their way with marvelously replaceable tanks of oxygen.

When the winter's first ice is only two or three inches thick, I have slipped up to a beaver lodge and jumped onto it, deliberately disturbing the tenants. I could hear a startled occupant splash into an underwater tunnel and could watch as it swam a hundred feet away from the lodge, emitting a trail of tiny white bubbles. Wherever it stopped, a large white bubble would suddenly form. Getting down off the lodge, I tiptoed out onto the windowpane; but the beaver never let me get closer to the newest bubble than a dozen feet. Then it left at a tangent and swam away, again leaving a tell-tale trail, until another large white bubble appeared. I have chased beaver around from bubble to bubble until, tiring of the game, they either re-entered their lodge, or gave me the slip under an overhanging bank.

Whether made by mink, muskrat, beaver, otter, or by methane gas, bubbles leave a pattern as individual as a fingerprint.

This woodsman's heart's desire would be to take a group of scientists, preferably chemists, on a field trip along the clear ice of any brook, river, pond, or lake. The scientists need be equipped only with open minds, strong legs, and winter underwear. I can find bubbles. I can tell who blew them. I can repeat the beaver experiment, and let the scientists do the chasing and the concluding. I yearn to understand what chemical miracle occurs during the life-giving moment when a water animal's nose "touches" ice.

Interpreting sign sometimes demands the use of "woods braille." On an abandoned tote road, a bunch of feathers lay on lightly crusted snow. Until the previous afternoon, they had belonged to a ruffed grouse. The snow was hard enough to hold up a fox; it would certainly have supported a weasel. But the hardy seed eater had been killed by neither. I picked up a primary feather to scrutinize the quill. Round and firm, it looked normal. But when my fingers moved along the hollow shell, they felt a clue, undiscernible to the eye. The faint indentation was the killer's mark.

Whenever a goshawk or an owl kills a partridge, it plucks out the primary and secondary, or wingtip, feathers. They interfere with its meal. To do this, it grabs hold of a quill with its powerful, built-in pincers, squeezes, and jerks. The force of the extraction flattens the quill, but only temporarily. Foxes and weasels also prefer plucked partridge, but, in pulling feathers, they puncture them with tooth marks. To be understood, some stories must be read with one's fingers.

Because life for the strong means death for the weak, many of the plays written by wild animals are tragedies. Even though a human brain gratefully accepts nature's immutable laws as logical and sound, a human heart cannot help but hurt for the helpless.

Before the freeze-up, the little marsh usually cradled five feet of water, but after four consecutive years of drought, the water table had dropped steadily, until barely eighteen inches covered the spongy bottom. Generations of muskrat had lived out their lives in the little bog. During late October and early November, working night and day, the last of these had built a typically commodious "split level" home. The spacious combination bed-dining-living room were on the upper floor above water. A roomy hall led underwater to the pantry: a marsh full of living water plants. The muskrat knew nothing about the drought. He did what muskrats had always done. How could he comprehend that what had been deep enough for grandfather, was not deep enough for him?

Thicker and thicker, the ice froze. One morning in February, Mr. Muskrat discovered his food was isolated in a deep freeze, for whose sealed door only spring had the combination. Driven by starvation, I could see where he gnawed his way up through the side of his house and slipped out into a fearful blast he had never felt before. A half-mile away, up a long hill was a small pond. Perhaps he could find an unfrozen hole along its shore through which he could slip. Resolutely, he left his home and his marsh.

Muskrat do not walk as smoothly as they swim. As ill at ease on land as humans are underwater, they do not advance in an uninterrupted straight line. Rather, they go off on sudden, sharp tangents, ten or twelve feet long, every twenty yards or so. The exile's erratic trail scalloped the snow for a hundred yards. When he reached a bank of plowed snow, he slid down into the road. It was covered with two inches of fresh powder, the finest canvas nature makes, snow so airy it would have shown the track of an angel. Without hesitating a moment, the fugitive turned west. He zigzagged up the road, climbing up and sliding down snowbanks, until his unique trail reached halfway to safety.

Meanwhile, reading the sign, I knew that a large mink had entered the swamp from the north. The fresh tracks, permeated with the tantalizing scent of steak on the ice, drove the marauder into the muskrat's abandoned house. Finding it empty, he began tracking his prey at a fast run. Electrified by visions of fillet, whenever he reached one of the muskrat's tangents, he ran beyond the scent until hunger shouted, "Slow down, you idiot, you're going too fast!" He would find the track, straighten it out, and taking jumps from twenty-four to thirty inches long, resume the chase at full throttle. In clear script, the headlong pursuit was written in the snow.

At high noon, on the crest of the hill, the mink had caught up to the muskrat. The lee of a snowbank was polka-dotted with blood. Bits of muskrat and mink fur lay strewn on the snow. Each desperate animal weighed in at about three pounds. Weakened by a more prolonged hunger, the muskrat lost. If it had been able to back up against a stone wall, the story might have ended differently.

The victor had to make a decision. If he left the carcass out in the open, his turn could well come at dusk from a great horned owl. The mink grabbed the bloody victim, and, moving backwards, began to drag it down the hill. Along the plowed road, he tugged and pulled the body back toward the marsh. At the foot of the hill, he turned into the bog and headed for an abandoned beaver lodge. He found a hole among the sticks and stones, then pulled his booty through it. For three delightful days, the mink had feasted on muskrat fillet.

As I studied the tragedy written in snow, I was torn by conflicting emotions: admiration for the pluck of the muskrat, pity for its suffering, wonder for the determination and derring-do of the mink. Only when driven by starvation or sex, will a wild animal risk a plowed road in broad daylight.

It always delights me to come upon a story that does not end in tragedy, but in a bruised ego. It had been a bitterly cold, open winter until a few days before, when a light snow had fallen. Followed by a sudden thaw, the snow had melted. Then a heavy rain began to fall. Because the frozen earth could not absorb the deluge, water formed puddles, some several yards long, in every natural pocket. Abruptly, winter regained control and rammed the thermostat down to zero. The falling rain turned to snow. Soon, three inches of powder camouflaged ice-covered pools.

Shortly after entering the woods, I came across a cat track made several hours before. The cat was traveling at its usual hunting pace: walking silently . . . freezing . . . waiting . . . as it zigzagged through woods and swamps. Steady, one paw in front of the other, the tracks finally led down a slight ridge. At the edge of an alder swamp, they came to a sudden stop. The cat's pace changed to nine-foot bounds. One . . . two . . . three . . . a turn to the right . . . a long, splayed mark ending in disaster.

While stalking the ridge, this cat had spotted a hare hopping slowly along the edge of the swamp thirty yards below. The cat froze, waiting until its dinner was fifty feet away. Bre'r Rabbit, suddenly realizing it was either quit eating or be eaten, shifted into high. In trying to intercept a fleeing dinner, the cat had to make an unplanned turn. The attack that had started from solid snow ended on ice. Tom's feet went out from under him as he slid, head over bobtail, for fifteen feet, to crash into some bushes. I could see where he got up, shook himself, let out a string of cat curses, and shaking a clenched paw toward the alder swamp, snarled, "Just wait 'til next time, Bunny Boy."

Few hunters learn their animal ABC's. As a guide, I have watched them tiptoe through the woods, rifles cocked, where there have been no deer for weeks. Any hunter who does not learn to read sign intelligently, will hunt forever in a forest of fantasy. His may even be the excited voice that stammers to a game warden, "The panther! I just saw the Black Panther!"

Each year, the advent of hunting season brings reports of sightings. A high school teacher earnestly told me he had seen one on a back road. (He teaches social studies.) Hunters invariably turn a "rehash" conversation to the panther. Leaning against a bar, one member tells how "it ran across the road in front of my headlights." With each round, it gets bigger and blacker.

In New Hampshire, as well as in her sister states, reports are received from natives, tourists, hikers, and from weekend nimrods, all of whom just saw the Black Panther. Rabbit hunters, bird hunters, deer hunters, all see it. A few years ago, one deer hunter got close enough to fire several shots, wounding it so severely that "the panther" had to spend three weeks in a local hospital bed licking his wounds.

After a weekend or two of deer hunting, most hunters don't set foot in the woods again until fishing season begins. Meanwhile, as they drive their cars over winter roads in remote towns, or skim along abandoned tote roads in their snowmobiles, their vehicles overwhelmed by snow-covered, forested mountains on either side, they can imagine almost anything skulking along those ridges.

While they were in their chariots, my snowshoes were crisscrossing snowbound townships daily from daylight to dark, in search of bobcat tracks. It was not unusual to search several days with perfect snow conditions, and not find a single one.

I understand cats. Hunters, they spend more time hunting

than I do. They roam a large area at will. Like me, they leave tracks in snow. I have never come across any track that remotely resembled that of a panther. Even if a track could fool me, the gait would not.

Why is it no one ever sees The Phantom when there is snow on the ground? Panthers do not hibernate. The deeper the snow, the lower the temperature, the bigger the appetite!

Did you ever eat wild beechnuts? I have yet to meet a woodsman who didn't enjoy nibbling on them. But one can try to assuage his hunger by eating them for hours, and still go away hungry. That is what a hare or two would be for a panther, an animal whose average weight is 125 pounds, with some known to reach 200. Such beasts cannot live on hors d'oeuvres. Scientific research has proved that panther boilers demand a deer a night. If there were a panther roaming New Hampshire's woodlands, the nightly dramas written would be bloodier than King Lear.

Tracks in snow are the easiest to see, but can be the hardest to decipher. A fresh fall of powdery snow in zero weather can preserve tracks exactly as a freezer preserves food. On the other hand, wet, slushy snow distorts any tracks made in it. Hunters have buttonholed me to sputter out a description of the monstrous deer track they had seen. Later the same day, I came across the Goliath and found it to have been made by a David. A small deer, running in wet slush, tells a tall tale. Running hooves, entering snow, force it to be pushed down and outwards beneath their driving momentum, not unlike you or I stomping our way through the same slush, leaving Paul Bunyan footprints.

One morning, I came across the largest human tracks I had ever seen in snow. Both my boots fit easily into a single imprint. Could Frankenstein be hunting in Cheshire County? Fascinated, I followed the trail. It didn't take long to catch up to the monster. Barely five and a half feet tall, he was wearing the first insulated rubber hunting boots I had ever seen.

4. *Woods' Mysteries: Secrets Nature Won't Share*

Nature's well-filled stacks hold many exciting mysteries. Those written in the woods do not revolve around "Who?" Rather, they make a reader ask "Why?" Then to puzzle "How?" Over the years, the ones I have read and pondered have filled me with wonder that has spilled over into faith.

Most encyclopedias and other naturalists claim that all foxes dig dens in the ground. This presents a riddle. During forty years of an intimate relationship with at least twenty-five "active" fox burrows, I have yet to see one freshly dug. How do foxes know where the dens are? During years when foxes are not plentiful, many dens display "For Rent" signs. In southern New Hampshire, most dens have not been occupied since the forties, when mange almost annihilated the fox population. Dead branches and leaves conceal deserted doorways.

Foxes do not hibernate, and, like most wild creatures, they sleep all day and hunt all night. When daylight sends their victuals scurrying for cover, foxes search out a flat stone in the open, facing the sun, wrap their fluffy tails, like boas, around their bodies, and curl up to dream of mice and men. They do not need to hide, for a fox's nose never sleeps. It is a built-in bodyguard, so sensitive it can smell man across 300 yards of forest on a windless day. I know of only three situations when foxes temporarily use a burrow.

In the spring, when she is ready to bear and to raise her young, a vixen searches out an old den, to use as a delivery room and nursery. After she has cleaned and "renovated" the place, she moves in. I have yet to see fresh sand around an occupied fox burrow, only litter and debris the wind may have blown in. When mother has inculcated the pups with all her foxy ways, she breaks up housekeeping. But, wherever they roam, foxes know where to find old burrows.

The second temporary occupancy comes after a heavy snowstorm. While waiting for sun and wind to settle snow, foxes will hole up until the snow is firm enough to support their nimble weight.

The third occasion comes when pressed too hard by a hound as smart and as fast as himself. That's when a fox, though an outsider in a township, will make a beeline for a burrow that may have been abandoned twenty fox-generations ago. How does it know where to find an ancestral den, dug decades before? With death at his heels, he doesn't speculate. He knows.

When he had reached his eighties, Arthur Leonard, who had hunted and trapped foxes all his life, had yet to see a freshly dug burrow. The holes my dogs chased foxes into, were the same holes into which his hounds had run foxes when he was fourteen, and the Battle at Appomattox had not yet been fought.

The ultimate satisfaction for a reader of mysteries is to resolve a riddle. In early spring, when the wind begins to smell of melting snow, and the forest floor to reappear, I find tufts of white hair resting on dead leaves. The varying hare has begun a biannual ritual which has piqued and defied scientist and layman alike for years. Twice a year, in a process that takes approximately five weeks in the wild, Mr. Hare changes his overcoat. A third molt occurs with no major color change, because it involves only his underwear. Like any sensible Yankee farmer or woodsman, his long johns are replaced in spring, augmented in winter. That isn't all. The hare has five toes. In summer, it can see between them as though they were buckhorn sights; but, while it is acquiring a winter coat, a heavy matting of fur is growing between its toes—not to keep its feet warm, but to enable it to skim over the surface of deep snow while foraging for food or while escaping enemies. Nature has strapped a pair of bearpaw snowshoes onto the hare's hind feet. In early March, when he begins to discard his heavy white ulster, his head turns brown first, his back next, his snowshoes last; but, when early in September, he begins to doff his summer coat, he puts on white snowshoes first, a white jacket next, and an ermine cap last of all. Why? What magic changes the sequence?

Recently, scientists have been able to artificially induce molting in the varying hare, by changing the daily amount of light to which the animals are exposed. It has been proved conclusively that the amount of light received by the hare's eyes triggers the change in color. Change is the key word. In other experiments, hare held under as much as eighteen constant hours of daily illumination forgot to molt. Up until this time, most people believed molting to be a natural result of weather and temperature. I never did. What shakes me to the soles of my rubber-bottomed, leather-topped boots, is not that scientists have been able to imitate nature, but the fact that the hare itself does not realize that it is white in winter, brown in summer. Hares are color-blind. Do you suppose these scientists think they have solved one of nature's mysteries? What would happen if they tried the same experiments with the red fox?

To me, the most baffling mystery of all is the navigational genius of an otter. A pup, orphaned before it has a chance to travel extensively with its parents, still knows, not only how to go from one place to another, but where he is along the way. Every otter inherits a built-in navigator and a compass. It's as though all plebes had been born knowing what admirals have to learn.

Shortly after the winter solstice, a pair of otter, while touring the Contoocook River, reached the headwaters of a minor tributary. From Woodward Brook, they watched the sun retreating behind Lovewell Mountain.

"Say, Sweetie," suggested the dog, "what do you say we go over to Halfmoon Pond? We haven't been there for a long time, and with all the camps closed, we should be safe. Besides, we've always had good fishing in Bog Brook. We should be able to reach it just about dark."

The pair swam north toward the outlet to tiny Ayers Pond, whimsically perched atop a 1,700-foot ridge. They were heading for a beaver pond they had visited a previous year in late autumn. There had been a prolonged drought that summer and fall, and they had had to pick their way over a dry streambed. It had been a rough trip. Otter hiking boots are not equipped with Vibram lug soles.

Since their last visit, they found the beaver pond had been abandoned. The water level had dropped. For lack of regular reinforcing, the dam had crumbled. Undismayed, the otter had gone under sagging ice to reach the inlet. Emerging from the tiny spring-brook, they set a northwest compass course, and took to the woods. They skied cross-country over a 1,600-foot-high ridge about a mile long. They had to cross an abandoned tote road, but, at twilight, they risked it. The isolated, overgrown road was unmarked by human sign. Usually, whenever otter approach humans, houses, or highways, they tarry, waiting for the cover of night before pushing on. No wise traveler moves through enemy territory by day.

Numerous slopes had provided the season's first exhilarating slalom and downhill runs. The last schuss had ended in Bog Brook, just in time for a big supper. Appetites appeased, they continued three miles down the brook into Halfmoon Pond. Now they were in the north branch of the Contoocook. Now they had a tripping choice. They could return to the main branch of the Contoocook via big Highland Lake and Jackman Reservoir, or the voyageurs could swim up Bog Brook to travel up the outlet to Frog Pond. A portage through the woods for less than a mile would take them to Bacon Pond, which is the headwaters of the Ashuelot River, a tributary of the big Connecticut.

If otter had not been designed with built-in compasses, consider the alternative journey: fifteen twisted miles backtracking down Woodward and Beards Brook to reach the Contoocook River on the outskirts of Hillsboro; a mile and a half to Jackman Reservoir; three miles to cross it; eight miles up North Branch River to reach the outlet to Island Pond; five miles up through Mill Village; eight miles plus up the length of Highland Lake; two and a half miles up the inlet into Halfmoon Pond: a forty-one-mile trip by water. How did they know?

Countless generations of otter have made that one-mile portage. When now-extinct birds were soaring over primeval forests, otter were using the same shortcuts. A thousand years from now, if we do not wipe them from the earth, future generations of otter will climb over hidden hills in the middle of the night to do what no civil engineer could do without schooling and instruments. Such a mystery can only lift one's eyes upward.

Sign to a professional woodsman is what a book is to a scholar. A scholar must be able to read and to understand words.

A good woodsman must learn to read and to comprehend sign. Both take time, thought, and tenacity. When a scholar goes to a library and picks out a book, he begins to read. When he cannot understand, he studies until he comprehends it. When a person who cannot read sign walks through the woods, it is as though he were walking through a foreign land. All he sees are trees, brush, and ground. He is like an illiterate who picks up a book, glances at it, thumbs through its pages, and discards it. Because the strange marks mean nothing to him, he might even contemplate a page upside down. History's tragedies, comedies, ironies, and mysteries are lost to him, because he has never studied enough to comprehend what he might read.

To a person who has learned, not only nature's mother tongue, but most of her dialects, a walk through the woods is like reading a great novel. Every few steps, new characters are introduced. The plot changes and develops; the tempo increases or subsides. Climax follows climax. One can hardly wait to turn the page. One walks only a short distance before uncovering new and exciting volumes in nature's library. All of them have "footnotes."

One does not need a library card to borrow nature's books. They are ours for the searching. Like all great literature, they touch the heart. Educating the heart stretches one's character much tighter than schooling the brain. In His acre, God is not dead. There are no atheists in the woods.

5. Untamed Affections: Compassion Among Wild Animals

I still remember the first wild animal I caught. It was a coney rabbit that had been strangled in my snare. The elation and pride of accomplishment, added to the knowledge that this creature would be a welcome and tasty addition to an otherwise meager supper table, struggled with an uneasy curiosity in my nine-year-old heart. What would the rabbit's mother say when he didn't come home for supper?

For many years, unanswered questions troubled me. Whenever I trapped a mink or an otter, or shot a fox, a bobcat, or a deer, I couldn't help but wonder how the creature's peers and family were being affected. Or were they? Whenever a member of my own family or a friend dies, I grieve in the same measure I have loved. How about wild animals? Do they love? Do they grieve? Four decades of observation has uncovered knowledge which I have never been able to find in any books.

In the woods, animals with full bellies also tend to have full hearts. The ones I know about are the beaver and the otter. The creatures who must scrounge for every mouthful to survive, especially during long and bitter winters, fight any competitor for the scant food supply, even in their immediate families. As living links in nature's food chain, all wild creatures live on borrowed time. The usury rate in nature's commissary is high.

How different are we from wild animals? Does appetite shape our thinking? Do we make the same choices whether our bellies are full or empty? Do well-fed humans always show compassion for those who are hungry, or love for those who have alien ideas?

We have split the atom. We breakfast in Boston and lunch in San Francisco. We orbit in outer space. Yet, in the most important area of all, we have feet and hearts of lead. What good to fly to the ends of the universe if we do not understand our place in it? What good to kneel in churches when we do not know how to love one another at home? What good to land on the moon while we still kill each other for power, for pleasure, and for profit?

In many ways, life among animals in the woods makes more sense than life in "civilization" among human beings. Do you suppose a wood mouse says, "Some of my best friends are meadow mice, *but*—"? Do you suppose starlings envy the scarlet tanager his fancy "mod" jacket? Do you suppose sparrows attack crows because they are black? Animal behavior is logical; wild creatures don't know how to rationalize.

Most wild creatures kill only to eat. Once they have eaten, they will not eat again until they are hungry. A bobcat, whose belly is full of varying hare, could come upon a bunny bebop session in a swamp he was passing through. He wouldn't give the dancers a second look.

A great horned owl sleeps all day. After dark, it sits up in a tree, its head turning continually, like a gun turret on a capital ship. It spots a hare hopping along a bunny trail, swoops down, scoops up its supper and returns to a tree. Holding onto a branch with one talon, it devours the animal down to the last hair and bone. Then it flies to another, bigger tree to sit and to burp. Along comes a whole band of bunnies. Their arch enemy ignores them. He is still burping varying hare.

Even those creatures who, in the merciless struggle to stay alive, fight their own kinfolk, maintain a sense of basic decency many human beings lack—or choose to ignore. Buck deer, dog foxes, wild tomcats, even domesticated male dogs show a moral sense many men are lacking. Animals do not fight their rivals to the death. The moment one of the contenders growls or bleats "uncle," the struggle ends. Knowing nothing about theology or religion, and without either the ability to reason or the conscience to feel guilt, most wild creatures fulfill God's plan more honestly than we do.

In the spring, when snow melts and the world turns green, wild creatures once more begin to eat regularly and without a struggle. A hundred different flavors of mice scamper across nature's smorgasbord. Insects and beetles grow with the grass. Wild strawberries stain fallow fields. In July, a sweet indigo carpet covers old pastures and forest openings. Underfoot, frogs, lizards, toads, and snakes hop and crawl. With the waning of summer, grasshoppers and crickets combine food with frolic. Choppings turn red and then black with raspberries and blackberries. In the woods, wild apples, acorns, and beechnuts roll underfoot.

Then winter sets in. Frozen apples and nuts are sealed under layers of frozen snow and ice. Mice huddle, hidden under the snow. No longer do little birds fall out of nests. No longer does the forest floor offer red-spotted newts as delectable tidbits. Berries and grasshoppers are only an agonizing memory. By midwinter, as much as four feet of snow buries the universal pantry. The thermometer shivers outside a north window. Sometimes, at night, it lurches down to thirty below. There is no central heating in animal homes. All heat must come from a little boiler room that can be stoked only through the esophagus. It cries constantly for fuel—any kind at all, just as long as it keeps the pilot light burning.

To stay alive, a fox must forage afar. Night after night, its tracks come out of the woods to a tree whose stiff fingers still grasp five or six rotted, frozen apples. He circles the tree. Perhaps a provident wind blew one of the bitter fruit to the ground. When he finds nothing, he trots off toward another tree he remembers several miles away. Night after night he returns. When an apple finally does fall, it may well be the first morsel of food he has tasted in more than two weeks. In mid-November, a mature dog fox weighs thirteen to fourteen pounds. His orange coat is glossy and a little tight over a paunch. Three months later, he is a bone rack. He barely tips the scales at five pounds. With several feet of snow under his still nimble feet, he sinks less than two inches. By February, the thread of life is stretched taut.

Among New England foxes, mating begins anytime after the first of January. It is usually during that month that their tracks first appear in pairs. Some naturalists claim that foxes are monogamous. If they are, theirs is not my idea of monogamy. The dog and the vixen have nothing to do with each other except during the rut and for a few brief weeks around the first of April, after the pups are born. Usually, snow still covers the ground, and food is hard to find. For a short time, papa hangs around the borrowed burrow. Once in a while, he halfheartedly helps with the groceries, but it is mama who carries the major burden. It is she who looks haggard. The old man is as dapper and as peppy as ever. Obviously, he eats his fill first. His devotion rarely lasts out a month. Then he abruptly shakes the pine needles of domesticity from his heels. It is the vixen who teaches the puppies how to be foxes. Is it she who teaches them never to share food, even while mating?

The rutting season for foxes had begun after a harsh winter when, week after week, the temperature had hovered around zero, and four feet of snow covered the mountainous landscape. I was snowshoeing along an old tote road on my way out of the woods. The sun had set. Already, darkness gripped the forest. Only the sky remained free. Suddenly, in the gloom ahead, appeared the shape of a large porcupine. Scratching noises followed as it climbed up a juicy hemlock tree for its supper of bark. I waited until it was silhouetted against the sky, shot it, and slung the carcass over a crotch in the tree high enough off the ground so that no dog could reach it. At daybreak, I was back following the trail where I had come out the night before. Today, the second leaf of my standard "three-leaf clover" pattern of hunting would unfurl.

During the night, a dog fox, accompanied by his lady friend, had come out of the forest to walk up my webbed trail. Even though the size of the tracks in the soft snow told me the owners were large, they were sinking less than three inches. When had these two had their last meal?

About a quarter-mile from yesterday's "porcupine tree," the foxes had smelled food. The tracks began to weave about, then to zigzag wildly. A hundred yards from the tree, they went berserk. Back and forth they ran, without either pattern or design. It looked as though a thousand foxes had been there. The hedgehog was gone. The tote road, including my snowshoe tracks which still had held my scent captive, was covered by a maze of tracks, clumps of hair, and tufts of quills. What perplexed me was that more fox hair was strewn about than were porcupine quills.

Savagely attacking each other, though it was during the rut, these two had fought over the hedgehog. The bloody battle continued for a hundred yards up the road. There, among a confusion of tracks, blood, fox hair, and broken quills lay a battle-frayed souvenir, the porcupine's most cherished possession, its tail. Completely encompassed with spears, even deadlier than those that armor its back, it is the ultimate weapon, its lethal power to be unleashed only when all else fails. A battered tail was all that remained of a twenty-pound animal. From the way that loaded weapon had been punted around, it was obvious the foxes had fought over trying to make both ends meat, even though they couldn't possibly have swallowed any of it. In the throes of courtship, even a selfish man or woman sometimes does unselfish things. Not so the fox!

Bobcats hate competition even more than foxes do. They tolerate one another only while they are little, and that tolerance never develops into affection. Like foxes, bobcats get together only during the rut. Even when driven by nature's strongest urge, a cat will not share its food. Once the female is impregnated, togetherness ends. A tom's role as a father lasts only for moments. (How like some people!) Under no circumstances, will a female allow a tom to come near a litter. No doubt, she knows him better than you or I! I believe that one of the reasons bobcats are not more plentiful is because, given the opportunity, the mature males kill and eat their own kittens. Under certain circumstances, they will even eat their peers.

One late February day, high up on Lovewell Mountain, I found an astonishing amount of cat sign concentrated in one small area. The tracks had all been made by the same cat. Why were some of them two weeks old? Why had others been made only seconds before my hound and I barged in? Why was Tom moving around in the middle of the day instead of sleeping?

What was he doing? Why had this particular cat forsaken its roaming ways to remain in this particular area for several weeks? Searching for answers, I climbed up onto a ledge, then followed a path leading from it into a thicket. The well-worn route had been made personally by the cat. As I emerged into a small opening, I almost stumbled over the frozen body of a bigger-than-average cat. It had been partially devoured.

In December, before the snows came, the frozen cat had been wounded by a deer hunter. It had crawled up onto the mountain to die. During the ensuing weeks, the continuous cycle of snow, freeze, and thaw had encased half of the carcass in solid ice. Another cat, while hunting the mountain for hare, had come upon its buddy in cold storage. It could even have been its brother. The parts not encased in ice made good eating, but as the meals progressed, so did the effort needed to get them. Finally, the cannibal had to tear at the remains with its claws as well as with its fangs. As I stood on the windswept ledge, my thoughts flew across a century and a continent to Donner Pass. The will to survive has burned as strongly in the hearts of some men as it does in the hearts of most cats.

A spiteful disposition seems to be as much a mink's heritage as is its sleek, shiny fur. In the spring, the female bears two to five young in an abandoned muskrat den, a natural hole in a river-bank, or, sometimes in an abandoned beaver lodge. Although most encyclopedias claim litters with as many as eight, I have yet to see more than five. No mink ever knows what it is to have a father, for mink are raised by their mothers without any help from the males. Ma herself has only a perfunctory maternal instinct, and abandons her children just as soon as they can fend for themselves. In midsummer, I have frequently seen three little mink from the same litter, hunting along a brook together. Mother Mink was nowhere around. By late summer the siblings themselves split up. Mink are loners. They share nothing. Rarely, and only under certain unusual circumstances, are they ever drawn together, and then only for a brief period.

The winter of 1933 was one of frostbite and chilblains, as any alumnus of the WPA or any woodsman could testify. During all of January, the thermometer never rose above zero, neither by night nor by day. Late in the fall, a small, isolated pond had been drained in anticipation of repairing a crumbling stone dam in the spring. When the pond had shrunk to its lowest level, the brook below the dam boiled with fish. The news ran quickly along the animal grapevine. Because, long ago, I had broken their code, the message reached me almost as quickly as it did seven mink. I arrived at the gold mine only a day after they did. (They were prospecting for fish.)

The assemblage had called a nervous truce and taken up temporary residence among the stones in the dam. Each had a private room with a picture window overlooking the restaurant. Love for food, rather than for one another was the common denominator. Whenever they weren't eating, they explored the corridors and nooks of their granite motel.

Among the large base stones along either side of the brook below the dam, I searched for places to set traps. Suddenly, my curiosity was excited by sign it had never seen before.

Moving backwards, a three-pound mink had dragged something much heavier than itself almost fifty feet from between two of the huge granite blocks. Several times, it had stopped to rest. My eyes followed the drag mark across the length of the dam and thence across an opening, up a slope, to disappear under a snow-covered brush pile. What had been worth such an unminklike struggle?

Removing my pack, I dropped to my knees and began to tear apart the pile of brush. To my amazement, I found a five-foot-long, forearm-thick, black water snake. It had been frozen solidly into the form of a shepherd's crook. The head and ten inches of the torso were missing. Tooth marks were etched into the last pink bite. The snake, hibernating in the old rock dam, had been discovered by a marauding mink. Needless to say, when the mink had made its find, it hadn't shouted, "Hey, fellas, come see what I found!" After stuffing itself in secret, it had buried its treasure like any pirate. To me, it was a big black water snake. To a mink, it had been a big black popsicle.

The whitetail deer has a gentle disposition, especially when compared to that of the fox, the mink, or the bobcat. But when that timid tail is pressed hard against the winnowing wall of winter, self-preservation prevails.

Doe drop their fawns in late May or in early June—one year, a single; twins, or even triplets, the next. Throughout the summer months, the doe cares devotedly for her little fawn. The latest I have seen a fawn nursing was in mid-September. During the fall, food is plentiful, varied, and delicious. The usual "meat and potatoes" diet of second growth twigs, buds, and berries of wild shrubs and red maples is supplemented with juicy wild apples, crunchy acorns, and toothsome beechnuts. Fawns remain with their mothers throughout autumn. It is then that doe teach their young how to avoid the thunderbolts who walk on two legs. If the young learn their lessons well, they survive, to enter a winter yard with their ma's. After three months of feasting at nature's harvest table, they are all fat and strong.

The congregating into yards begins when snow reaches a depth that makes roaming about in search of food too difficult. The largest single yard I have seen in New Hampshire was in Stoddard. On a mid-February day ten years ago, I watched twenty-two deer moving about listlessly, after surviving a harsh winter. (On the tenth anniversary of that day, I revisited the area. Not a single deer had passed over the dozen miles of forest I covered on snowshoes. There was no deer yard. There were no deer, even though there was as much browse as there had been a decade ago.) Deer do not yard up just anywhere. They try to pick an area that has a healthy growth of both deciduous and coniferous trees. Until spring, they browse on the deciduous branches and buds, and find shelter among the spruce, pine, and hemlock thickets. Though deer do not enjoy browsing on pine and on spruce, they do relish hemlock as well as balsam twigs.

Many times, I have passed through yards and found the inhabitants on a starvation diet, eating pine needles and pine branches. Less than a mile away, ample, nourishing food awaited. Why didn't the deer go find it?

During many years of snowshoeing into many yards, to approach within a few feet of starving deer, I asked myself this question. Why don't these pathetic creatures want to leave the well-worn trails of their winter prison to find forage elsewhere? Even though starving, they remain. The longer they remain, the less there is to eat. The less they eat, the weaker they become. Finally, malnutrition so dulls their mental and physical faculties that they allow a mortal enemy to come almost within touching distance.

In summer, fall, and early winter, when deer are in top-notch condition, the flick of a white tail is all most people ever see. During hunting season, the majority of shots are aimed at a rapidly disappearing north end going south. Now, in a browsed-off yard, they stand still. With glassy eyes, they stare at my approaching figure. Deer have a keen sense of smell, and man wears the wrong kind of perfume. Closer and closer move my snowshoes. The deer does not know whether to run or to stand still. Finally, when it sees the whites of my eyes, it takes a few feeble bounds, stops in plain sight, and resumes staring.

Suppose a vigorous man had been handling a responsible, decision-making job exceedingly well. Suddenly he is put into a tiny room. Once a day, a bowl of cornflakes is placed before him. For the first month, there is milk and sugar also, then only sugar, and, finally, only a meager half-cup of the dry flakes. After two and a half months, the door suddenly opens. His boss has a dangerous problem that demands an instant decision. Would the "prisoner" be able to grasp, to think, and to move as quickly as he had before his internment?

In a winter yard, the doe, who was so solicitous during summer and fall, does not help her young to stay alive. When browse can be reached only by the biggest deer standing on their hind feet, mature doe do not break off branches for the young who cannot reach and for the weak. In the woods, life is a personal responsibility. In some eastern states, conservation departments have tried to feed starving deer by introducing fodder into winter yards. Immediately, the big bucks took possession. What they couldn't eat, they befouled. The doe and the yearlings didn't get a cudful. The weak starved. The bucks, because of their size and strength, would have survived anyway. Even though we sometimes think we have caught up to nature, we can't outwit her. Where nature, rather than human nature, is in control, there need be no fear of population explosions.

Even though I have never seen compassion among yarded deer, where, as a human, I felt it was needed most, several moving experiences have convinced me that some deer are capable of caring. The first occurred in the thirties, a few days before deer season opened. I was running a trap line and, as always, carried a shotgun loaded with birdshot. I was hiking along an old tote road that separated two ponds. The dams had long ago been washed out. From the old causeway's high, dry surface, I could see almost a half-mile around and ahead. Among the speckled alder and the chest-high marsh grass that had repossessed the pond beds towered great, gray boulders. What an ideal place to spot wild animals!

My eyes scanned the farther end of the basin first. Scarcely fifty yards from me, with only their heads showing above the grass, stood two deer. Sensing me, the royal pair bounded away. A massive, twelve-point buck led his slender, almost girlish queen up the basin, and away from danger. Gracefully, he leaped first to the right and then to the left, his sweetie close on his heels.

About a hundred yards from me stood a gray-black, bungalow-sized boulder. Its steep, slanted front faced toward me. It was almost like a movie screen. The springing buck approached the rock broadside. Without slackening his pace, he passed in front of it, his magnificent bulk silhouetted against the stone, to partly disappear among some alders. The doe took two bounds, then stopped broadside against the granite backdrop. Her attention no longer centered on her lord and master. It was riveted on me. Female curiosity had conquered feminine obedience. What was there to be afraid of, anyway?

Suddenly, the buck realized that his mate was no longer following him. He spun around to face her. He looked at her; he looked at me; he twitched his tail; he looked back at her. For tense moments, his eyes moved back and forth from me to the queen of his heart. Every glance spoke clearly in universal, masculine tones. Abruptly, with tremendous bounds, he leaped forward to disappear behind the boulder. Before I could blink, he reappeared on the other side. Still at full gallop, he ran behind the doe, lowered his trophy-sized antlers, and jerked them up with all his outraged strength. Her hind quarters flipped a foot and a half off the ground. She flew up the basin, the boss right on her tail. That punt in the rump had been in her language. ·

The young doe had obviously never been shot at, but the old buck interpreted the situation differently. A veteran of many hunting seasons, he no doubt still carried buckshot under his hide. To him, any man with a gun meant only one thing. I have never forgotten that big buck who cared enough for his doe to risk his life for her.

Another time, a middle-aged doe unforgettably demonstrated her ability to love. While guiding two deer hunters, I came upon the fresh tracks of three deer. Those of a large adult intermingled with those of two smaller ones: no doubt, a doe with her twin yearlings. After being "jumped," they took off on the run toward the men waiting on stands a quarter of a mile away.

Even when shot at, unless wounded, deer rarely double back. So, moving fast, I angled away from the direct line of the tracks. In case there was a miss, I could intercept the escaping deer, turn them, and give the hunters a second chance to fill out their deer tags. Upon hearing a fusillade of shots, I stopped underneath a stand of tall hemlocks.

Suddenly, less than fifty yards from me, the fleeing doe put on her brakes with such force that leaves and debris splattered her heaving undersides. She was not a young deer. She looked full at me, hesitated for a moment, then turned to face the direction from which she had just fled. Ignoring me, she paced back and forth, her great white flag rising and falling, her big, upright ears twitching forward and back. Liquid, dark eyes were turned hopefully toward the scene of the shooting. Across the clearing that separated us, her anguish reached out and touched me.

A mixture of emotions rushed through my heart. Because of me, both her yearlings were dead. At that moment, I would have given almost anything if I could have undone what had just been done. I knew she had seen me. I knew she was terribly afraid of me, but she loved her twins even more than she feared me. If she had been the last deer I could ever shoot, I would not have raised my rifle.

Anyone who studies beaver cannot help but be impressed and improved. The beaver's engineering skills, its single-minded devotion to duty, its perseverance in the face of unnatural as well as natural obstacles, surpasses the liveliest imagination. Their ability for solid group cooperation can make philosophers of us all. But there's a muscle-pulling step between mutual acceptance and loving response.

Poppa Beaver doesn't throw his arm around Junior's shoulders and say, "Son, that was a good day's work you did tonight. I appreciate it." He hasn't the time. Besides, that kind of display might well loosen the tight reins needed to maintain discipline within the colony. A communal society usually demands rigid behavior.

In every lodge, a motto hangs over the door: "He who will not work, cannot eat." No beaver ever forgets it, not even the littlest one. (Daddy doesn't have that broad, flat tail for nothing.) A beaver's life of toil makes the salt mines seem appealing. I have never seen a beaver play. For them, life is just one dam after another.

I cannot agree with those naturalists who claim that some beaver prefer to live alone. By heritage and by habit, these woods engineers possess a group fidelity as compelling as that of the ant, the bee, or the Daughters of the American Revolution. Beaver need each other. How could one engineer construct a dam, build a lodge, dig canals, and store a winter's food? I have yet to see a beaver with battle scars. They do not fight among themselves. Dedicated wholeheartedly to the same cause, they do not waste their energies and their characters fighting one another. Their ability to work and to live together in natural harmony can only command our awe and our respect. But, like Paradise itself, their social order has flaws, especially when viewed through man's moral bifocals. Within the colony, the individual is nothing, except as he benefits the group. The colony always comes first. Because they live such interdependent lives, no beaver leaves a colony by choice. He departs only by decree.

Eighteen years ago, halfway through the beaver trapping season, I was snowshoeing up a frozen deadwater to reach an old beaver pond in hopes that it had been resettled. About a mile below the dam, the river narrowed, then twisted down a series of short pitches over a tumble of rocks. Even in zero weather, large openings among the churning rapids remained unfrozen. As I was detouring around the white water, a fresh trail in the snow surprised me. They were the tracks of a solitary, mature beaver. To find an unmarried beaver is unusual enough. To find a celibate who is also "eating out" in the middle of February is a phenomenon. What had happened to this production engineer's time standards?

It had emerged from the fast water to make its way into the woods. The tracks led to a birch tree eight inches in diameter. It had been cut down a week before. Since then, each night, the beaver had emerged from the white water to eat his fill of bark. Something was wrong. Beaver do not usually eat their bark sandwiches on the stump; it's too dangerous. Mr. Flat Tail could himself end up as a hot lunch for a bobcat. After filling up on birch bark, the maverick re-entered the water and swam upriver. Why was this beaver breaking all rules of beaver behavior? Who ever heard of an impulsive engineer? Pondering the paradox, I snowshoed up to the pond.

The lodge had been abandoned five years before. This loner had been living there since early autumn. Never before had I seen a bachelor beaver. Never before had I seen a beaver who left a lodge in the winter. Never before had I seen a beaver who didn't have a feed pile. Never before had I seen a beaver that swam a mile for a meal. What had happened? Why was he doing these unnatural things? Only the beaver could answer my questions.

As a professional, I set a trap so a wild creature caught in it dies quickly. The best place to catch a water animal is in its own element where it cannot get to shore. With beaver, this means trapping them underneath the ice. When a beaver or an otter is in danger, it heads for deep water as naturally as a drowning man struggles to reach the surface. When trapped properly, an animal drowns quickly in deep water.

Fifty feet below the lodge, I located a suitable place in the pond, where this recluse would have to pass, en route to his unfinished birch dinner. After clearing away the snow and chopping a hole in the ice, I set a trap. To it was attached a stout piece of a dead tree. Beaver do not cut dead trees, only live ones; they like tree sap gravy with their stakes. After baiting the trap with a piece of freshly cut poplar, I lowered it into six feet of water, and covered the hole up with snow. The next time I made my rounds, I cut out the mushily frozen snow, lay prone on the ice, shielded my eyes, and stared into the murky water. The trap was gone. The chain stretched out of sight downstream. I had my beaver.

I lifted the dead pole onto the ice, grabbed the chain, and pulled the beaver out of the hole. Even though it was an immense animal, weighing at least sixty-five pounds, it was not a patriarch. No gray silvered its muzzle. It was in the prime of life. Then I saw why it was living alone, why it had no feedpile, why it had to travel so far to get its food. This beaver could no longer store food. Both of its front feet had been amputated. The brutal surgery had been performed a year before by novice trappers.

How could this beaver carry mud and sticks for a dam? How could he help fill the family pantry? What good was he to the group without his front feet? When this beaver could no longer lift his end of the log, he had been expelled from the colony.

It was not unusual for me to catch beaver with one foreleg missing. Like most other states, New Hampshire is interested primarily in revenue from trapping licenses. The state does not require trappers to furnish evidence of their ability to trap intelligently and humanely. Most part-time trappers, armed with a one-dollar mail order booklet on "How to Trap," and a burning desire to get rich in one season, head for the culverts. Trapping is a profession. It takes years of thoughtful toil to acquire the knowhow that makes it profitable for the trapper and comparatively painless for the trapped. After an initial eye-opening season, most of the would-be millionaires give up, but not before leaving unnecessary cripples in our rivers and our lakes.

Since that day, I have caught several "bachelor" beaver. These were not crippled. Each was a gigantic male. Each was a "senior citizen," certainly a great-grandfather—perhaps even a fourth- or fifth-generation patriarch. Why was he living alone? you ask. Here is my theory.

When beaver mate, they swim off to find a likely brook, build a dam and a lodge. They start a family. After the first litter of one to four kits, Mrs. Beaver usually has quintuplets. Naturally, father and mother are boss. No matter how tempted, teen-age beaver hesitate to question either the authority or the judgment of their parents; it's pretty hard to pierce armor that always practices what it preaches. In their third year, the children, now newly graduated engineers, leave, pick a mate, and get married. Like some people, a few of them come back to the old homestead to live. Trouble brews. Here are the parents still telling their grown-up children what to do. Sometimes, when grandchildren arrive, the emotional stew begins to boil. It is a struggle to raise children, even without interference. "Well," bristles Grandpa, "that's not the way *you* were raised. I swear, I don't know what this generation is coming to." As Grandpa gets older, his advice gets stuffier.

When a similar impasse occurs among humans, the old folks are banished into nursing homes. As yet, beaver do not have "rest homes," so the troublemakers are expelled from the colony, to live out their remaining years in exile. For some strange reason, I have never found an old matriarch living alone. Obviously, beaver lodges have only father-in-law problems.

If God has a favorite animal, surely it is the otter. Why else would he have been so generous? As if the gift of a built-in compass wasn't enough, He gave the otter the gift of love. I believe otter are monogamous because a lifetime's experience with their character, personality, and habits has convinced me that for them it's "till death do us part." Whenever sign tells me that a single otter has passed through my territory, it is to me a definite indication that this otter has lost its mate, not by divorce, but by trap or by gun.

Twenty years ago, in the dead of winter, a group of them clearly demonstrated their ability to love. I was snowshoeing up an isolated river, ten miles long. Even though this was ideal otter country, I was amazed. Never before had I seen the sign of nine or ten at the same time and place.

They were not just passing through. They had spent several weeks fishing and swimming together along the ten-mile stretch of river. They fished the frozen deadwaters and romped in the ice-free rapids. Sometimes, they split up into pairs, with some of them going upriver while others went downstream for a few miles. But they always regathered. The sign showed they had been there for several weeks.

I have often noted three, or even four otter, from different watersheds, together for a week or two. Never before, or since, have I observed so large a clan gathered for such a long reunion.

Because otter love each other, the family circle is a close one. Food and fun are shared. Whenever death strikes its hard blow, the bereaved otter do what no other animal except man does: they mourn. I can remember the first otter I caught. I picked up my trap with a never before felt sense of accomplishment. Within a few years, I discovered that whenever I trapped one of these beautiful animals, its mate and its family grieved. Otter truly feel grief. There is no other animal I know of, once its mate is trapped, that will remain in the vicinity, sometimes for many weeks, searching and waiting for its beloved. It seems willing to risk a similar fate to find out whether it is really true that its loved one is gone forever.

After thirty years it still hurts me to remember a family of four who were fishing a forty-mile section of the Ware River and its tributaries. Studying their particular timetable, I concluded they would return to the starting place in about two weeks.

The upriver journey always ended at a remote backwoods pond which nurtured not only several varieties of fish, but also an unusual abundance of crayfish. I knew how welcome these would be to any otter. At the appropriate time my traps were set in the terminal pond. As expected, they remained empty until the seventh day when two held drowned otter. One was a female yearling, the other an exceptionally large dog, whose size and worn yellow teeth indicated he had "commodored" many a convoy. Three days passed before the third otter, a yearling male, was caught. Then nothing happened for more than a week.

Could there have been only three? Could the mother have succumbed to a hazard other than myself? I thought not, but it troubled me. Something just wasn't right. One dark, overcast, late November day, when the air was so angry it spit snow, I hiked up to the pond to see whether a solution to the mystery could be found. From a tangled clump of laurel I could view several hundred yards of water in three directions. Within a half-hour I observed what I had come to learn. First some ripples several hundred feet from shore, then a disturbance close to the surface, followed by a familiar flat head. It was the fourth otter. Not only her head but her entire body floated on top of the water, something I had never seen before. She was neither fishing nor playing. Motionless, she riveted her attention on the south shore as though expecting something. After a few minutes she turned to gaze in the opposite direction.

Her peculiar behavior continued for twenty-five minutes before she slowly slipped under the surface, leaving only a few gentle ripples. I watched from my hiding place for another half-hour, expecting her to reappear. She did not. I wanted to wait longer, but the bitter cold forced me to leave.

For the first time in my life I had seen a forlorn otter. It gave me an unfamiliar feeling that clawed at me—deep down. At the time I didn't try to analyze either her strange behavior or my stranger emotion. It took many more years of wooded and domestic maturing before I understood what had happened. Now I know that this widow had searched in every distant cove of the pond, repeating her sad ritual as she grieved for her loved ones. They had suddenly gone out of her life because of me.

Since that dark day I have confirmed the existence of this strong bond in otter families many times. I know what happens at every final parting, even though I am not there to see it. The bereaved one no longer plays. The lively dives, the explosive emergings, the lusty snorting and throat-clearing, the animated head-jerking, are all things of the past. The whole attitude has changed to one of lassitude. They dive slowly, emerge listlessly. During the long, sad weeks of searching, of waiting, and of hoping, widowed otter seem bewildered.

In 1947 I was trapping one of the main arteries of the Contoocook River, where three otter were periodically visiting a certain two-mile stretch of the stream. After the usual waiting period, the skins of two of them, one an adult male, the other a male yearling, were stretched on my boards. During the next two weeks I made several "dry runs" in my canoe to the remaining traps. Where was the third otter? Why wasn't she still on the river? Somewhere on that two-mile stretch of riverbank she had left a memo for me.

Determined to read it, I swapped my paddle for my legs and headed upstream through the woods for the part of the river where towering ledges forced the water through a narrow, rock-filled gorge before releasing it to plunge into the large, deep pool that formed the upper end of a deadwater. I broke out of the dense spruces onto the ledges above the fast water and looked down into the pool. Something moved, but too far away to identify. I had to get closer. My approach was hidden by a heavy screen of tenacious evergreens. Parting the branches of a spruce, I searched the pool's surface.

An otter's head appeared, but without either the usual splash or a fish. I was close enough to distinguish its long gray whiskers and silvery throat. Its shiny dark eyes looked steadily toward the arc of white water splashing down into the farther end. Then rolling slowly underwater it reappeared, facing downstream, to gaze unwaveringly toward the vacant river. I knew it was waiting for someone to suddenly swim upstream or to lunge over that wall of rushing water.

There was only a ragged remnant of satisfaction in my ability as a trapper when I realized the otter was crying. Whenever it surfaced, it let out a series of four or five throbbing, plaintive, almost dovelike sounds. To me those noises were not those of an animal. I had never heard this sad lament before, nor have I heard it since. But how often does a woodsman find himself where he can see and hear the anguish of this gentle creature's breaking heart?

I got my last otter in 1957. During those last years of my intensive professional trapping I could no longer feel the triumph I had once felt. Upon removing a drowned otter with its broken teeth and limp, exhausted body from my trap, I could think only of the mate and its fruitless search for the lost one. Then I felt compelled to catch the survivor, more to end my own misery than theirs.

Yes, I had indeed mastered my art. Now that art had boomeranged. What had started as a battle of wits developed into a partially self-supporting way of studying a fascinating creature in its own element. It ended in an almost passionate desire to protect the animal to whom God has given so many special qualities.

Would-be trappers have offered me hundreds of dollars to teach them how to catch otter. I refuse them all, not because I don't need the money, but because I have learned to love the otter. I would feel like a traitor. This much I am willing to pass along. It's the same answer I gave a game warden when he asked me how I knew where to set an otter trap.

"I set it where the otter's going to put its foot."

The knowledge of how to do this will remain mine alone. It will go with me to my grave. My chosen profession has served me well. The animals were good teachers. The otter were the best of all. Only by being an otter myself would I be able to understand more.

6. Yankee Bobcats:
Valiant, Variable Vagabonds

In late 1945, after this sailor swapped his uniform for a hunting
shirt, he found that mange among foxes had wiped out a former
source of winter income. Dame Fashion had also worked hand in
hand with Mother Nature. Long-haired furs were no longer in
style. So, because the state paid a bounty of twenty dollars for
each bobcat turned into a local conservation officer, I had a clear

and solid indication of my need to learn how to hunt them. The unthinkable alternative was to go into a factory. With the combination of trapping, guiding, and an adequate bounty from cats, I could stay in the woods and keep the household kitty solvent.

The first thing I learned was that gunpowder, two legs, and a half-domesticated brain barely balanced a wildcat's four legs, eagle eyes, and radar ears. The bobcat's eyes do more than see, and its ears do more than hear; for its eyes and ears are a cat's "brains." They warn it of danger. They find it its food.

Hunting cats is a grueling and lonely task. From dawn till dark, day after day, week after week, for more than three months, the one human voice I heard was my own. There was only my dog to talk to, or a cat to curse. By the end of March, when the snow began to melt, my clothes hung on my frame. The best we had as a man and a dog was pitted against the best they had as cats.

They fascinate me. Of all our domesticated animals, the cat is the most independent, the most individual, and the most regally stubborn. Its wild cousin shares all these characteristics. Wary, unpredictable loners, I never so much as caught a glimpse of any bobcat, unless it was driven by one of my dogs. Even after a winter of near starvation, it is still a beautiful creature: head like a miniature tiger, eyes tawny and luminous, spirit unceasingly heroic, whiskers curving out in long, stiff arcs.

The second lesson I learned was the big advantage a cat has in the kind of environment it chooses. When trailed by a hound, a red fox utilizes open country, stone walls, fields, roads, barways, and even a half-mile of railroad track to make good his escape. A bobcat does the exact opposite. His trail is a heart-stretching obstacle course. When forced to cross a road, cats pick a section bordered on either side by outcroppings of ledges, or better yet, wherever a spruce swamp barely allows a road to dissect it. When going cross-country, they choose the swampiest swamps, the rockiest ridges, the thorniest thickets. Between mountain ranges, they go through every tangled blowdown. Do you want to undertake the pursuit?

Another early discovery was that cats know more about New England weather than either I or the almanac does. During a series of clear, cold days, I have scouted ledges in search of cats. Sign showed that none had been around for days, sometimes not for weeks. On my way out of the woods, the clouds were so low I almost carried them on my shoulders. The wind shifted to the northeast. Before daybreak, the howling of a blizzard would awaken me. Smiling self-indulgently, I would turn over and go back to sleep. Two days later, after the storm had abated, I would return to the same ledges. Within hours after my departure, the rock pile that had been unvisited for days had been occupied. An hour earlier, a cat had emerged to venture abroad to test the snow. Finding it too soft and deep for hunting, it had retreated to the cave and like me, two mornings earlier, continued snoozing. The next morning, the exercise was repeated, and the same conclusion reached. On the third morning, hunger grabbed the reins, and drove the cat into the swamp at the foot of the mountain, even though the snow was still too deep. That's where studying his habits made it possible for me to catch up with him.

Bobcats do not hole up during snowfalls of two, four, or even eight inches. But when a major storm is brewing, something tells them, "You've got about three hours before the blizzard hits. The snow will be too soft and deep for you to hunt. You might as well sleep it out."

Over and over again in a dozen ledges on a dozen mountains, I have seen this ability to predict the amount of snowfall repeated. How do cats know? I can only assume they are equipped with something better than a barometer.

Another thing I was not long in finding out was the great advantage of a pickup truck over a car. I had known it before, though not about cats. People wondered why a woodsman like myself drove a pickup truck. To them, it didn't make sense. I wasn't a farmer, a contractor, or a carpenter. Why a truck? The answer is logical. All our wild animals, except deer, have fleas. (The deer hair is so coarse, it would be like living in a broom.) An eye-to-eye visit with any squirrel, fox, or rabbit will reveal a startling number of parasites scurrying cross-country along the clearings around the eyes and nostrils of their hosts. Even mink and muskrat harbor an occasional wood tick, although I have yet to see a parasite on a beaver or an otter. The breed of fleas that inhabit bobcat country is especially large and voracious. Without actually having counted them, I believe every cat supports at least two thousand of these lively boarders, whose size is matched only by their bite. Cat tracks tell me they flop down regularly to scratch, scattering clumps of hair as mute testimony to an irresistible itch. Many times I have come out of the woods dragging two cats. I soon learned that tossing four thousand fleas into an enclosed car was a mistake. Throwing their hosts into the back of a cold, drafty pickup truck made the ride home more comfortable. At least, I could keep both hands on the wheel.

Yankee cats come in two colors: cinnamon brown with lots of yellow, or storm-cloud gray with very little yellow. Both have underlying darker stripes. Neither sex, size, nor environment determine these variations in color. Nor does color affect their social lives. Theirs is a desegregated society, and mixed marriages are common.

The average mature cat weighs about sixteen pounds. They vary in weight as much as we do, though not for the same reasons. I have shot some weighing as little as seven pounds and some as much as forty. Unlike us, a cat's age can usually be judged by its weight and its teeth. The older it is, the more it weighs. Old teeth are always yellow and worn.

Bobcats are selfish, independent, and mean. They are only sociable for a few weeks during the latter part of February and the bright days of early March. Deep inside some glacial rock pile, the young are born during the first two weeks of May. Most encyclopedias say there are from one to five kittens in a litter. I have never come across more than three. Even a trio is rare. But I know the hazards every young cat must face. Not the least of these is a mature tom who will kill any kitten, blood relative or not, astray from the safety of its mother's claws.

When the young are big enough to trail along, mother cat teaches them to hunt. This schooling is thorough, and classes last until late fall. The trouble is, like much of our education, it's all theory and no practice. One autumn night, mother cat brings in a rabbit. Snarling, she tosses it in front of her kittens. They are playing and squabbling among themselves as they have done since early summer. Startled at her strange behavior, they look at one another.

"What do you suppose got into Ma?" asks one.

"Maybe she got up on the wrong side of the den," answers another.

"Aw, she'll get over it," ventures the third.

They don't realize she won't get over it. Ma has had it. She just cut the gastronomic umbilical cord. The next night there is no rabbit, partridge, or squirrel on the table; nor on the second or the third nights. Every night, for as long as three weeks, the bewildered, starving kittens, hoping for a handout, follow their mother at a safe, respectful distance. Only a snarl and a clout are forthcoming. Ever increasing hunger finally warns them it's find their own food or starve.

The abandoned litter hunts together, often until the middle of January, which is the latest I have observed. It was usually a pair or a single kitten whose tracks caused my hound's sensitized nose to twitch. This doesn't point to no brothers or sisters, but rather to the fact that life is a hazard. After the bum's rush suffered from Ma in late autumn, food is still plentiful. The swamps abound with inexperienced rabbits. Wild apple trees bend with fruit. Bustling chipmunks and squirrels grow careless. Only a few hunters roam the woods.

Then comes the deer season. Overnight the countryside is aflame with nimrods, standing on rocks, thrashing through swamps and blowdowns. On rare occasions the deer hunter, motionless on a stand, with a shamrock in his pocket and with a battalion of other hunters in the area stirring up sleeping animals, spots a fleeing pair of cats and bags one. Only a cat as lucky as that hunter, can elude so crowded a gauntlet.

Cats are nocturnal hunters. This includes house cats. Well-fed, pampered, hopefully spayed, but instinct is inoperable. When night falls, tails switch and eyes glitter, whether it be by the hearth or in the woods. Wildcats usually hunt from dusk to dawn. Then, oftentimes, they find a sheltered ledge that faces toward the southeast, and curl up for a good day's sleep—until rudely interrupted by a hound dog on their scent. On dark and stormy days, like other denizens of the night, cats tend to move about more freely. In the early spring their habits change abruptly. Driven by a force more gnawing than hunger, they become quite reckless as, during the day, they roam about searching for a mate.

Born vagabonds, mature wildcats range many miles in their daily travels. After some two thousand snowshoe miles following their tracks, I can usually tell what the animal is up to. I don't always know why. A beaver lives in its lodge, a rabbit hops its lifetime away within a single swamp, unhunted deer rarely browse out of a three-mile circle, but cats have a hunting route that sprawls over as many as five townships. A given cat will hunt the same swamps, visit the same ledges, and sleep in the same motels every time he's in town. Cold tracks or hot, they all lead up over ridges and through deer yards searching for a winter-weakened deer or a foolish young porcupine who is still slow on the draw. Cats never attack a full-grown, live "porky." Because each respects the other's defense system, they just "dip the colors" when passing. While hunting over a twenty- or thirty-mile circuit, cats oftentimes hole up in dens located in ledges or glacial rock piles that are already infested with porcupines. But an apartment teeming with tenants doesn't faze a cat. Forcing his way in, he snarls, "Push over. You've got yourself a guest for the day whether you like it or not!" The patient porcupines move over, but not because they are afraid. They know the balance of power rests in their tails.

When "jumped" by a dog, a cat can be as unpredictable as a panic-stricken human. If a cat heads for the nearest tree or rock pile, it is usually either very young or very full. Inexperienced kitties have more confidence in their climbing equipment than in their senses, and any cat that has been stuffing itself with deer steak has developed a weight problem. How can it run when its pot belly is dragging on the snow?

Some cats are draft dodgers. They want no part of any conflict, especially one that is thrust upon them. In order to avoid bumping into the draft board, they quit hunting long before daylight to duck into a convenient hideaway. I once tried unsuccessfully to beat a tom like this to his retreat for five consecutive mornings until, on the last day, while still too dark to shoot, I stood defeated at his front door two miles up a mountain.

His direct opposite seems to relish nothing better than a hot chase. These Hairbreadth Harries play in front of a hound better than a snowshoe hare, zigzagging merrily around and around within a quarter-mile circle, always in either a swamp or in a thick blowdown. When the chase reaches the boiling point, the confident quarry leaps up on top of fallen trees to run there until it catches its second wind, while temporarily outwitted hounds cast around for a fast-fading scent. Such reckless toms have dashed by me with only seven feet separating a bobbed tail from a dripping enemy tongue.

The least profitable bobcat to hunt is a steeplechaser. This type loves to run cross-country and, when jumped at daybreak, sets its sights on the horizon. The champion led me through seven townships and back again until, five days and a punishing fifty-mile round trip later, the marathon ended where it had begun. Working sixty hours portal to portal for twenty dollars is only for a man who loves his work.

The most select cat is the Old Smoothie. This sophisticate has discovered that few dogs can dance to his tune. Veteran of many a ball, he heads for a favorite mountaintop ten or twelve miles away. There he picks out a ballroom where he "trips the light fantastic" ahead of his partner while the firepower weaves his way on snowshoes through the snarled warp of the woods. The wisest old customer I ever saw was a big tom who approached a ledge below me. Well out of shotgun range, he casually sat down in front of a readily accessible motel door to listen to two hounds pounding on his trail a half-hour away. Finally he got up, stretched, yawned, and sauntered into his bedroom for a well-earned catnap.

Although their tracks seem to zigzag aimlessly, cats do not meander. They know where they are going and how to get there. One of their peculiarities, while traveling in deep snow, is the way cats take advantage of any available means to keep from sinking. Suppose there is a stand of large spruce trees fifteen or twenty feet apart, whose branches hang close to the snow. The cat will climb onto the outer end of a suitable branch, claw its way to the tree trunk, and then continue its chosen direction by tightrope walking down another propitious branch on the opposite side. Thence on to the next tree via mountain monorail.

Any cat will follow the top of a long windrow of brush left by loggers in a chopped-out area. Naturally, for many years after the 1938 hurricane, they took advantage of the blowdowns covering thousands of acres of mountain slopes. Whenever I approached a blowdown of forty-foot spruces, bare of small branches, and lying tip to trunk, the plural cat and dog tracks would end. The hound's tracks would go on alone. After walking on the fallen trees for more than a quarter-mile, I could still look down and see only his trail weaving through the brush. When the blowdown ended, the track of the cat would suddenly reappear. Like me, it had crossed on the tree trunks, taking the elevated, while the dog struggled in the subway, his nose telling him only that the other traveler was somewhere above and ahead.

In ordinary travel the measured step of a cat is approximately eighteen inches. Whenever snow is more than four inches deep, Tom has a tendency to slide his paws in at an angle, quite unlike a fox, who puts one paw directly ahead of the other. In deep snow, where only the indentation and not the pads can be seen, a large cat track resembles the hoofprint of a deer, partly because their strides are similar. When danger threatens, we run faster, leap farther, and climb higher. Cats' react the same way.

Like all wild creatures, bobcats are unique and consistent in their personal habits. They usually defecate near ledges, atop boulders, and along the high land of abandoned tote roads. The droppings consist of bones and fur, an occasional broken or bent porcupine quill or two, the fur of both red or gray squirrels, and sometimes, during the winter months, the coarse hair of a deer. A wildcat has many of the same instincts as his half-domesticated relative, particularly in the area of hiding the evidence. It tries to cover its scat with disorderly swipes which are naturally more casual than those of its more cultured kin. I once met a bobcat who considered a more sophisticated approach.

But to begin at the beginning. It was still dark when I edged my rubber-bottomed, leather-topped boots into the resisting leather of snowshoe harnesses, stiff with cold. Leaning over to fasten the buckles, my unmittened fingers were begging to be housed again. It wasn't going to be a day for dilly-dallying. I had barely limbered up when I came across the spoor of a large cat. Like many tracks, it had been made in the early part of the previous evening. It was hours old. Should I take it? The size of the pad helped me decide. My four-legged business associate approved wholeheartedly and together we began to "work out" the trail.

"Working out" is a literal phrase when applied to tracking a cat, especially when, like this one, it led over one steep mountain range, several dense, snow-neck-filling swamps, and too many trouser- and temper-tearing blowdowns. The sun had slipped behind Nancy Mountain when the trace finally zigzagged into a small clearing behind an isolated camp on the shores of a for-

saken pond. Seeming to sense that the premises had been abandoned for several months, my bewhiskered will-o'-the-wisp made
its way through a small thicket and, as if by intent, came up to
the camp's outhouse. For some reason the latch was up and the
door partly open. The winter's snows had built up around the
building, and a dusting of snow had been blown onto the floor.
The tracks approached the door with no apparent questioning.
They showed it had been opened wider and the facilities scrutinized with crude excitement, before the animal regained its
dignity and went on its way. It was worth the twenty-mile excur-
sion on snowshoes to find out that a man-made two-holer didn't
quite fit Tom's standards.

In the area of oral communication, wildcats are oftentimes
given credit for speeches they don't deliver. Rabbit hunters tell
me they have heard bobcats snarling in swamps. Some farmers
insist they have heard bobcats screaming behind their barns at
night. The nocturnal vocalizing is real enough, but it emanates
from the throat of a fox. From the spring melting to the fall ice, I
have heard their yapping countless times. I have heard it summer
nights while training fox hounds. I have heard it on a frosty
evening while pout fishing. Most people associate the verb "yapping" with barking. Both words are woefully inadequate when
describing the goose-bump-raising sound made by a fox at night.
It is a scream a maiden in dire distress could use or a banshee
could boast about.

I am particularly intrigued by the claims made by responsible
biologists and naturalists, who assert that wildcats prowl around
the woods yowling and screaming. Surely, eyes must twinkle
when the statement is made that cats squall while stalking a
quarry in order to paralyze it with terror. Have you ever watched
a house cat stalk a chipmunk or a bird? Silent as a snake, it
slithers forward on its stomach, and Tabby isn't even hungry.
Consider its wild cousin. For half of each year, nature locks the
door to her larder and buries the key under several feet of ice

and snow. Food is hard to find. Keeping one's belly full is a bond we share with wild animals.

Suppose you were lost in the wilderness. It has snowed for three days. You have not eaten in six. When the storm abates, you begin to stalk a deer. Would you shout, "Look out, up ahead, here I come! I'm going to kill you!"

Accepted authorities on nature have written that, when mating, wildcats screech and scream. Thousands of hours of my life have been devoted to studying the bobcat. I have spent many months, of many years, hunting it in its own element. I have yet to hear a sound from a cat that was free, whether sexually stimulated or not. I believe the cat's mating is compulsive, quick, and silent.

I remember an early March day after a prolonged period of bitterly cold weather when the rutting season was at its height. When I had entered the woods at daybreak, the dry, frozen snow had carried on a conversation with my snowshoes for the seven miles up onto Willard Mountain. By ten o'clock, the sky clouded over and the temperature spurted upward. Within an hour, the snow was soft and porous. While descending the northwest slope I came across cat tracks of various sizes made since the sudden thaw set in. A closer examination revealed that a quartet was traveling together. During the rut, one naturally expects to see two bobcats together, occasionally even a trio, but never four. Experience and common sense told me it must be a single female being courted by three males. The tracks were only minutes old. We were in a large chopping, covered with long windrows of snow-covered brush from which occasional dead branches protruded. Threatening black clouds rolled over the peak like cannonballs. Suddenly, the world was small and dark. What an opportunity for a sighting!

For nearly an hour I stalked the quartet. I am sure they heard me, but they did not panic, nor did they head in any definite direction. Around and around we zigzagged, many times so close to one another that I could almost smell cat. I never caught a glimpse of one. The only sound was that of my own breathing. How I wish the cats had squalled!

Trapped or wounded, and forced to fight for its life, a cat will snarl and spit while making savage swipes at its tormentor. Under natural conditions, though they are many things, bobcats are not vocal. I welcome an invitation from anyone who has observed a squalling cat. I am willing to spend several midwinter nights in a sleeping bag to be proved wrong or right. It seems to me that too many assumptions are made about wild animal behavior. Or could it be a matter of geography? Influenced by a dour environment, the New England Yankee tends to be a taciturn cuss. Do you suppose it could apply to New Hampshire cats as well?

All wild animals, except the beaver and the otter, have a real struggle to survive during the long winter months. Unrelenting, constant cold triggers thermostats. Hard-pressed boilers demand more and more fuel. Most of summer's snack bars are closed. Several feet of snow covers the communal pantry. Cats have an especially rough time. The gullible young partridges and rabbits have grown up; the chipmunks are sleeping and the woodchucks hibernating; the varying hare is harder and harder to find.

The varying hare, commonly called the snowshoe rabbit, is the wildcat's chief source of food. Cats get a rabbit dinner only by surprise, never by pursuit. If hunting conditions are ideal, they average one hare a night. They eat it, head and all, leaving only the last two inches of the hind feet. Evidently, they don't like the snowshoe harnesses. Tom's menu also includes red and gray squirrels. Cats do not catch these creatures in trees, but stalk them as they do hare at their usual hunting pace: walk silently . . . freeze . . . wait . . . steady . . . one paw in front of the other . . . stop . . . freeze . . . wait . . . bound, bound, clutch.

Occasionally, before the freeze-up, a cat will pick up a teenage beaver cutting logs on a side hill away from the pond. In midwinter, cats often torture themselves by climbing onto the melted south cones of beaver lodges, to stare down, blunt noses pressed against frozen mud and sticks, at the delicious tenants below.

The bobcat also fancies porcupine fillet and, when pressed for food, will kill a small one. Someone else has to do the dirty work where the big ones are concerned. Any man with hunting dogs has a personal vendetta with all porcupines. They already owe him at least the cost of one trip to a veterinarian's. So he shoots any "porky" on sight. A cat will never pass by a dead one, but will approach these free dinners carefully. After turning the carcass over onto its pincushion back, it tackles the soft underside with gusto, eating all but the hide and quills. I often find quills deeply imbedded around a cat's mouth. Like me, they have learned the hard way that no one gets something for nothing.

Many sportsmen believe that the bobcat menaces our deer herd. This is not true, even though they enjoy deer meat as much as I do. On the contrary, as charter members of nature's Survival of the Fittest Program, they help to maintain a healthy herd. Full-grown, healthy deer may be half-starved, but can still keep out of Big Kitty's reach if they don't do anything foolish. Evidence of such foolishness is seldom found in cat scat. When coming near deer that have not yet yarded up for the winter, a cat's hunting walk will suddenly change into eight- and nine-foot bounds. Anywhere from five to a dozen leaps will end at a still steaming deer bed. Have you ever watched "sleeping" deer through binoculars? They tuck their lean legs under their bodies, which remain motionless. Not so the head. Rarely down for more than a few seconds, it jerks up to scan every inch of its bedroom. A startled deer takes leaps fifteen feet long. It lifts up with a thrust that rockets it into action. A marauding cat finds only the still hot takeoff pad, enveloped in clouds of rotting leaves and snow.

On rare occasions a good-sized cat does pull down a deer, but only during the bitter end of a bad winter. By then the whitetails have lived in their yards for almost two months and have browsed all the available food, even to the point of standing up on their hind legs to reach higher branches. Because size limits the reach, the yearling deer suffer the most. Malnutrition weakens them mentally as well as physically. Panicked by a prowling cat, they will sometimes leave the comparative safety of the yard, where the snow has been trodden down. To leap into the deeper snow beyond is to wallow helplessly. Whether walking on melting crust or on four inches of fresh powder atop three feet of packed snow, a cat can improvise snowshoes by spreading out its toes; whereas, betrayed by its hooves, the enfeebled deer is belly-deep in agony the moment it leaves the yard. Now the cat has the upper paw.

After the kill, a cat gorges itself on the fresh meat. A good-sized tom will eat six or more pounds of deer meat in a single night. Following the banquet, it straightens out the dining room by getting on top of the carcass to scrape snow, hair, and blood over the messy remains. It thinks it has done a perfect cleanup job so that no one else can recognize its meat storage. As far as other cats are concerned, the camouflage is perfect. From seventy-five yards away on an open, windswept ridge, I have identified one of their caches: white snow, red blood, brown hair. The evidence convinces me that bobcats, like all other furred creatures, are color-blind.

After dinner, the gourmet searches out a close-by rock shelf facing the sun, to sleep there in true tiger fashion. It is "eat and sleep" in delightful rotation until only the deer's chassis remains. Then the cat will resume its roaming. But it always revisits the skeleton. The memory lingers on as Big Kitty regnaws a skull, picked clean several weeks before.

Cats never dine together. I have found deer on which two had been feeding. The sign told me they were from the same litter, but they never ate supper at the same time. The snow revealed the Cain and Abel struggle. The stronger one couldn't stand to share a steak, even with its brother. "Cain" gorged himself while "Abel" hungrily paced around and around just outside the steak house. Hunger sometimes overcame fear long enough for "Abel" to attempt a lunge and a quick bite. With exacting peevishness, the bully drove him away. Finally, glutted into a stupor, "Cain" will curl up on a nearby ledge to sleep. The waiting "Abel" soon learns that only time and intemperance are his allies.

For bobcats, a happy by-product of deer hunting season is gut piles. They find frozen deer plumbing irresistible. To cats, these visceral mounds, strung out through the woods, are a chain of Howard Johnson's, recommended by Tabby Hines.

Because most wildcats hate to get wet, they will take a long detour to avoid even crossing a brook. They swim only when forced to choose between life and a big, black hound hot on their tail.

It had been an unusually cold day in the Somerset country of southern Vermont. All morning I had been snowshoeing hard, with the double-barreled purpose of finding a track and of keeping warm. The results on either score were meager. Ahead of me, my four-legged business associate jerked up his head, then thrust down his nose. The snow was pockmarked with prints. Obviously, more than one cat had been responsible for the "fox and geese" design. From all directions the tracks converged into a balsam thicket. Even before pushing aside the sharp needles, I knew what was there. Just as in New Hampshire, all roads led to Ye Olde Steak House.

The chase was as short as the cat's breath; within twenty minutes Jiggs was barking "treed." Stretched out on the broken-off top of a tall dead spruce, a snarling cat glared down at its enemies. The shotgun spoke and the dead cat's wide downward arc ended on the snow. Jiggs leaped to grab his prize. I hung the limp body in a crotch of a nearby tree. Its rival still lurked in the vicinity, so we made our way back up to the cafeteria. The state of Vermont pays a ten dollar bounty. In order to maintain my minimum wage, both cats were necessary.

Within minutes, the dog "jumped" the other cat. Judging by the way it took off, it must have been "number two" at dinner. Running in a line drawn straight by panic, the cat tumbled pell-mell down a mountainside. My snowshoes substituted for skis to slide down the steepest dropoffs. When the baying signified "half-treed," curiosity spurred my stride. The dog was running back and forth on what appeared to be the edge of a cliff. Edging head and shoulders over a sheer cornice of solid snow five feet high, I stared down at the black current of the Middle Branch of the Deerfield River. It was like looking into a huge vat full of boiling ink. Three miles above, at Somerset Reservoir, tons of water gushed out from under the dam, generating a current too swift to freeze. But where was the cat? Then I saw it.

Sometime in the past, a large tree had toppled into the river and the angry current had swiveled it diagonally downstream. The frantic bobcat had leaped onto the bobbing trunk, scrambled out to the very last branches, and while still forty feet from the opposite shore, it dove into the icy water. From the end of the fallen tree the current carried the cat seventy feet downstream before it could clamber up the opposite bank to safety.

The hound's big body quivered as he waited for my "Get him!" Next to hunting, he liked swimming best. He was neither as agile, as light, nor as well equipped with claws as his adversary. He might drown trying to climb that opposite bank.

"Good Kitty Guy," I said, turning away and patting his brainy black head. "Not today . . . maybe tomorrow."

7. *Such Hounds,*
Such Headaches!

Hounds are the workhorses of dogdom. They put in long hours, endure rugged conditions, expend great amounts of energy, often experience frustration. The bobcat hound works longest and hardest of them all. He must contend with snow, ice, blowdowns, swamps, exhaustion, and sometimes with boredom as well as excitement. When it snows, he doesn't roll over and go back to sleep. He's up and ready before his master's belt is buckled.

85

When Mr. Average Hunter buys a hound, he thinks his dog is perfect—just like his kids.

"John, I got a helluva good cat dog!"

"That so? How many cats did you get last winter?"

"Two." (With a dog like that, a professional hunter would starve.)

For some reason, few men can face the truth about their hound dogs. They may confess disappointment in the woman they married, or in the children they fathered. Never in a hunting dog.

As a hound-dog man, I have trained many, cussed a few, and loved and respected two: Tim and Jiggs. You can always tell a good hound. Like a good man, he earns his keep. He knows what he's supposed to do—and does a little more. In many ways, dogs are like people. There are many breeds, many colors, with every one a surprise package—or a Pandora's box.

I know hound dogs as some men know women. Experience has taught me that a good man-dog team is more rare than a happy marriage, because it is as difficult to find the right dog as it is the right woman. No real hunter ever sells a superior dog. When one offers for sale a six-year-old dog, "the best I've ever had," he's either a liar or a fool—probably the former. Money can't separate the exceptional man-dog team any more than it can break up a good man-woman pair. The real thing is unbuyable, unbreakable, unbeatable. It ends only when one of them dies.

Buying a dog from a magazine advertisement runs the same risks as investing in a mail-order bride. She will be the color and about the age you specified. The impression you got from her photograph was fairly accurate. But can she adjust to you? Can you work together? Will you know how to take care of her?

Your mail-order hound will arrive with long ears, of the breed you ordered, all the necessary shots for distemper, rabies, etc., and perhaps a pedigree that traces his ancestral tree back to the acorn. Don't be misled. Can he learn? Does he want to work? Do you know how to teach him?

I pick a dog by his eyes. If calm clear eyes meet mine, I speak. Even though I don't know his name, the dog looks steadily at me, wags his tail, and waits for me to say more. He pays attention. Shouting is not necessary. Even though still a puppy, he doesn't jump all over me, or dash around in circles. He has the inborn self-respect and dignity of a little child. Sudden, unusual sounds don't send him scuttling, tail between his legs. If a car backfires, he heads for the sound instead of the cellar.

If we picked presidents by looks, we'd have missed Abraham Lincoln. The same holds true for dogs, for ability cannot be judged by looks. Beauty can boomerang when you least expect it. Whether or not a dog is registered matters little to me. I've owned hounds with names like racehorses, and with price tags to match. Heredity plays the same black-sheep tricks on dogs it does on people. Skeletons rattle in the best of kennels as well as in castle closets. Like a good man, a good dog must have courage and character. Blue blood and blue ribbons guarantee neither.

Any hound has more to start with than the would-be hunter. Dogs are born knowing much a man must learn. On the other hand, to assume that all hounds are born hunters is like assuming that all men are born doers. Koffka showed that if a kitten is ever to become an able mouser, it must learn in a given short span of its growing up. Perhaps this is equally true of dogs and people. Be that as it may, to assume that all dogs will become good hunters is like assuming all men will develop into good producers.

We know that good people do not always make good parents. Similarly, able hunters are not always adequate teachers. Fiber and basic personality come sealed in at the factory. Either it is inside the package or it isn't. Then comes the question of what environment does to the raw material.

A real dog man takes the task of being a teacher as seriously as a Dewey or a Casals. He knows it is easier to teach good, new habits than to break bad, old ones. His pupil hunts only where, how, and when the "prof" dictates. He would no more allow his charge to roam the woods at will, than he would thrust his little children, unattended, into the streets of a ghetto. Many a bright dog has become a delinquent for lack of a master who cared.

Caring is a demanding occupation. The more talented the pupil, the more knowledgeable the teacher he needs. A gifted child cannot develop its full potential unless its shining qualities are recognized. Neither can a first-rate dog rise above a second-rate teacher. Too often I have met a man whose dog had all the makings of an extraordinary hunter: the nose, the heart, the brains were all there. Alas, the same could not be said of the man on the other end of the leash.

A good master needs above-average intelligence; a comprehensive, personally acquired knowledge of wildlife; a sense of humor; and a few courses in psychology. A full-blooded vocabulary also comes in handy. He must know how to apply what he has to offer to the dog's natural abilities. To over- or to underestimate is to invite disaster. To hunt is a hound's birthright; to direct, a master's responsibility. A hound supplies instinct; a master, guidance. A sound man-dog relationship must be a two-way street paved with mutual respect and mutual need. A hound furnishes the means; a master, the end—and vice versa. It takes a goodly part of a man's lifetime to acquire the knowledge and the know-how, whereas hounds born with the right combination of nose, brains, and heart are as rare as comets. That's why so few of these wonderful man-dog teams course the countryside.

Some dogs are stupid. They never learn the lessons of experience. Ask any porcupine. Some are lazy.

Some dogs are rugged individualists. Unafraid to stand up when everyone else is sitting down, they scorn the pack. Unwilling to run at anyone's heels, they never run for the sake of running. They always know what they are chasing and why.

Bugle-voiced, with warm topaz eyes, Tim was one of those. He cared only about me and hunting foxes (I'm not sure in which order). He figured we made a tough team without any bolstering. He was a smart dog.

Some dogs are joiners. They can't stand doing anything alone. Upon hearing a pack of mouthy strangers several miles distant, this type will leave a fresh but lonely track, to join the bandwagon as it howls through the woods. The noise and excitement are what count. What's being chased isn't important. If accidentally separated from the pack, he is lost. Because he has never made an independent decision, he doesn't know how to find his way back. This gutless character is the backbone of any pack. He thinks with his mouth.

Some dogs are drifters. Unreliable creatures of impulse, all the training in the world can't change them. They rarely finish anything they start. After a dozen frustrations you are tempted to shoot them. Then a perfect run sends your hopes soaring. Happily you trudge home exulting, "Now he knows!" The next morning the newly-snowed-on mountains are a cat hunter's dream that borders on the idyllic. You stare down at a track less than an hour old. Off dashes your "new" hound, baying like a bloodhound about to corner its quarry. That's the last you see of the reformed character until hours after dark, when he casually reappears to greet his half-frozen master with a "Gosh, where-have-*you*-been-all-day?" look in his artless eyes.

Some dogs are politicians. Handsomer than most, this type is expert at lobbying. He has learned which side his bone is marrowed on. Humbly, he licks your hand; eagerly, he jumps into a car; impatiently, that perfect muzzle sniffs at spoor. After finding a still steaming track, you let him go. Three days later, a dog constable in the next county calls you. A charmer, this type means well. He starts out wanting to do what is expected of him, but it's those delightful diversions along the way he can't resist.

Some dogs are stubborn. Too opinionated to learn, they make the same mistakes over and over again. Whenever this type runs game into a burrow, it digs and howls around the Alamo for hours, until dragged away on a leash. Three miles of rough going later, when released, he makes a beeline back to the fort. The pathetic part is that stubborn animals often have good minds, but something goes wrong on the assembly line, so that they think with their hearts and feel with their brains.

Some dogs are phonies. As with people, it's easy to mistake size for endurance. A big, rugged hound, with rangy legs built for speed, seems to have all the qualities needed for clambering over snow-covered ledges at top speed. Straining on a leash, he dances about on his hind legs, yowling, "Let me at 'em!" Unleashed in the woods, he makes a half-dozen senseless orbits before falling contentedly into step behind you. He's tired. His muscles are all in his tongue. He's saving his energy for his next meal.

Some dogs are late bloomers. Time after time, while training him, you try to keep a young dog on a leash in step directly behind you, as you follow an old track, mile after mile. Every time you go over a log, the dog goes under, and when it's easier for you to go under, he goes over. As you go on one side of a sapling, he goes around the other, all the while still on a leash. The struggle of wills ends only when the tracks become fresh enough to make the dog sniff, then strain on his leash. You let him go. Within minutes a hot chase ensues, but something seems peculiar. He's not going anywhere; he's running in circles! The day is ruined because Tom's trail happened to wander too close to a bunny's bedroom, and once they take a track, most hounds forget they have a master. One trains and trains such a dog. Sometimes, after six or seven years, it develops into a respectable hound, but it has only a year left to hunt. It's like people who struggle all their lives to get good at something, to no avail, and then suddenly, just before they die, they blossom.

Once in a million litters, a genius is born, a happenstance as rare among dogs as it is among people. Being a blueblood has little to do with it. Neither Leonardo da Vinci nor William Shakespeare was produced by selective breeding. A genius hound has many talents. He is as loving a pet as he is valiant a hunter. Smart as a circus dog, he acquires a repertoire of tricks which he eagerly shows off on command. His nose can distinguish any smell. Whether it runs, crawls, or flies, he can identify any creature by its scent. But he follows only that for which he has been trained. His brain and his nose are connected by a street called Clear Thinking. He has more than a superior brain; he has a superior mind, and he uses it to observe, to compare, and to remember. Like an army veteran who understands the whole, almost second-nature routine of action, a good hound comes running whenever you pick up a shotgun. A genius will help you to lace up your boots.

By some strange magic he respects what his master knows, senses what his master feels, trusts what his master tells him. That is, if he has a competent and kind master. Otherwise, he is disgusted and shows it.

As a bounty hunter, I depended on a dog to help me to support my family. Many hours of my life went into training individual bobcat hounds. In order to do it right, one can train only a single dog at a time. Among other things, one needs a leash, lungs like bellows, and luck. On snow, you find a cat track, usually several hours old. Under normal conditions, fox or cat scent permeates tracks for about an hour after the animal has passed. With your pupil on a leash, you start following the trail, moving as fast as possible, because you must get close enough to your quarry so that some body scent lingers. The cat is not sitting down waiting for you to catch up. Unless it stops to eat or to sleep, or, unless you can outrun it, the spoor stays cold.

Only a rare dog ever learns to identify tracks by sight. Few learn, by themselves, to follow a specific track, ignoring all others, no matter how alluring. Whenever a creature other than the one you are hunting crosses the trail, and you never before realized how numerous they can be, you emphasize their unpopularity by a painful jerk of the leash whenever a muzzle yields to temptation, accompanied by the stinging rebuke of a switch over and over again.

If your stamina and luck go hand in hand, you'll get close enough to warn or to "jump" the cat. Unsnapping the swivel, you command your pupil to "get him." Don't expect him to—the first time—or the second, or even the third or the fourth time. Each time he will make mistakes, and it's your job to help him understand what these are. If he's as smart as you are, he'll learn from his errors, and rarely make the same ones twice. Finally, the delightful day will dawn when patience, persistence, and pluck are rewarded. Once a good hound has sunk his fangs into the warm, odiferous, flea-ridden carcass of a cat, he matriculates. Within a couple of seasons, if he works like a dog, he should be a graduate of Canine College. If he's like Ol' Kitty Guy, by the time he's three, he'll have an honorary degree. So will you.

Ol' Kitty Guy had perception. He soon learned that cats are unpredictable and hard to get. He not only grasped what I wanted, but what he had to contribute in order to get it done. He understood that one cat differed from another. He neither relied on any rule of paw, nor took his quarry for granted. Like all geniuses, he had superb self-discipline. Straddling spoor, still as a statue, ears cocked, he would wait for me. While I studied the sign, asking myself where this cat came from, where it was now, where it was going, how far it was to the nearest rock pile offering cat refuge, Jiggs would never even blink. He was all attention, waiting my command.

Sometimes, my judgment said it would be a useless pursuit. I would pat his black head and say, "Not today, Kitty Guy. Maybe tomorrow." If, on the other hand, the track looked workable, two words were spit out. Excitement, zest, and confidence put the accent on the first syllable.

"Get him!"

Wh-o-o-o-o-sh. He would disappear in a spray of snow. No matter how hot the track, he always waited for my command.

My dog must be as much of an individual as I am. He must be as sensitive as a baby, yet as tough as a combat Marine. He needs to know when to coo and when to charge. My hound never gives up—unless and until I do. Then, it is always a truce, never a surrender. My dog understands my thinking. He understands my hound dog dialect. After a wrong move on my part, a cat would outmaneuver us by running into a ledge. I'd peer into the hole. "Well, Kitty Guy, that was a bum move I made. The son-of-a-bobcat gave us the slip."

The dog would look up at me, then peer into the ledge.

"Never mind, ol' man," he'd wag back, bumping his head against my leg. "Not today. Maybe tomorrow."

Talking to my dog has shortened long miles trudged through the woods after dark. To me, a good dog is a companion, a friend, a confidante to whom I can tell things I wouldn't tell a psychiatrist. Whether or not my hound dogs understood, they all responded. As of now, I speak English Walker, Redbone, and Black-and-Tan. Who knows, maybe I still have time to learn Bluetick, Plott, and Leopard Cur.

8. A "Genius" Hound Comes to Stay

It was one of those February days when going to work was impossible in my profession. I turned from looking at the frost-bound kitchen windows to giving attention to my white-and-yellow hound. He lay with his head on his paws. I bent over to examine one. He licked my hand and I scratched a special place behind his ear.

"Well, Skivvie, it's napping behind the stove for you. Yesterday's hunt cost you too much."

Raw, bloody, his pads had no skin. If only his feet matched his pluck! I had raised him from a shivering, terrified-by-a-two-hundred-mile-train-ride pup, a dog whose blood was as blue as it was expensive. Stubborn as witchgrass, he served the longest apprenticeship in my hunting history. Now, at seven, he had blossomed into a better than average cat hound—except for those tender feet. I wondered if being a blueblood had anything to do with it. Who ever heard of a mongrel with delicate feet?

As a bounty hunter, I couldn't afford to take a week off because my partner's pads hurt; I needed another dog. The slow search for a young hound that suited me must begin. If I bought a six-months-old, registered dog, chances are he would arrive with a good nose, strong legs, and a bay my neighbors would soon complain about. But papers wouldn't guarantee brains and common sense. A brainy dog is like a brainy man: how much he's got isn't as important as knowing how to use what he has. My memory backtracked over the forty odd hounds I had owned. Most of them weren't worth remembering. I let them disappear behind the first rise of forgetfulness, but one dog stopped to stand in the middle of my mind's eye.

Rangy, cinnamon brown, and barrel-chested, with paws like a young lion's, Tim had cost me fifteen dollars. Hound-dog men eat macaroni and beans for weeks in order to pay ten times that for a top hound with a fancy name and a family tree that might put the royal Stuarts to shame. Where this yellow-eyed, wrinkle-faced, redbone orphan originally came from remains a mystery, even though I spent as many hours as dollars trying to unravel it.

We were two of a kind, for Tim disliked hunting with other hounds as much as I disliked hunting with other men. A pack hunter must have a pack personality. My grain runs the other way. I taught each of my hounds to hunt one specific animal. This private tutoring was wasted on some of the students. They never graduated anyway. But Tim was a Rhodes scholar. He remained a dedicated maverick throughout his life.

Year after year, we matched characters and dovetailed personalities while hunting together from the first of October until the first of March. He resented being left behind during November, while I ran a trapline, as well as during the December deer hunting season. Each evening, when I went out to his kennel to feed him, his tail would wag, but golden eyes reproached me. He thought I was chasing foxes alone.

During our second season as a team, another subject was added to his curriculum. On sunny October mornings, our fox hunts evaporated with the frost, so I taught him to hunt for game birds by scent. He was an apt pupil, but any bird hunters I took out never believed they could shoot their bag limits behind a bawling hound dog until they had seen big Tim in action. He even learned to retrieve partridge, woodcock, and pheasant. He picked up a duck once, to spit it out immediately. Then, coughing and sneezing, he lowered his tainted muzzle to the ground, and pushed it around in the marsh grass as though he were pushing a vacuum cleaner.

Hound dogs are not bred for "soft" mouths. Their silver tongues are all business, and Tim was all hound. Because he wanted the tasty wildfowl for himself, every bird resulted in a mad dash, followed by a tug-of-war. He was just naturally possessive.

In those days, a red fox jacket was to a woman's ego what a mutation mink is today. Whenever a fur dealer arrived to do business, the fox pelts were brought out of storage and laid on the kitchen floor. There the buyer would examine each one before starting to haggle. Tim had seen neither hair nor hide of his victims since he had shaken their limp, warm bodies. With a "so-that's-where-you-are" look on his corrugated face, he would jump up from his special spot in the living room and come over to smell each of his dead enemies. When the dealer leaned over to pick one up, Tim would straddle the orange-red pelts and snarl, his yellow eyes glittering. Those foxes were his. He allowed only me to touch his treasures; he did concede they were half mine.

After Tim died, several hounds came into my home, but none came into my heart. If a hound man owns one exceptional dog during his lifetime, how can he ask for another? I had had mine. The telephone rang my mind back into the warm kitchen. It was a local conservation officer.

"John, do you want a young hound dog? There's one for the taking up on Black Mountain. An old fella' who trains bird dogs for a living moved into the abandoned Baker place last fall. I met him on the road today and he asked if I knew anyone who would take a hound dog off his hands. It seems a nineteen-year-old boy he knows up in Claremont just got drafted. The kid brought the hound with him from Kentucky, somewhere up in the mountains. He thought too much of his dog to have it destroyed. Three weeks ago, he gave it to old Corson. You'd better get right up there if you want him. He can't keep that hound with his bird dogs much longer, and he doesn't want to shoot him unless he has to."

I still don't understand the sense of urgency that compelled me. I usually take plenty of time to make a decision, especially about a hound, despite which I hastily laced up my worn boots.

"Well, Skivvie," I said, "looks like you're going to have a buddy. Don't know for how long. I don't even know whether this hound comes from hunting stock."

My truck was left where a snow plow had given up in despair. A five-mile snowshoe hike into a stinging storm ended at an old homestead half-buried in snow. Several feet away from the sagging house, on a low knoll, a ramshackle barn rattled its skeleton in the wind.

Lean, wiry, and walking lightly on his toes, Mr. Corson looked like a middle-aged bird dog himself. I liked him at once. While putting on his boots, he wryly admitted that a lifetime in Arkansas had prepared neither him nor his wife, a full-blooded

Sioux Indian, for the rigors of a New England winter, in a fully ventilated farmhouse, on the north side of a mountain. We wallowed through the drifts to the barn. Five healthy pointers were chained to separate posts. Off to one side stood a tall, full-chested, black hound with a penguinlike waistcoat and white spats on each paw. His muzzle was October oak-leaf brown, as were the quarter-sized spots over each eye. His glossy fur seemed as dense as an otter's. He looked like a hound. Hope and misgiving clashed head on. His not having "papers" didn't matter, but was he a hunter? Had he sprung from hunting stock? Should I take a chance? Well, I never knew where Tim had come from either.

The dog was handsome, no doubt of that, but what struck me was his dignity. Only ten months old, he stood quietly, his head high, his tail up, not wagging wildly as one would expect in so young a dog. Appraising brown eyes searched appraising blue ones. A dog carries its character, as well as its heart, in its eyes. The dark eyes, looking steadfastly into mine, were calm, honest, and bright with intelligence. But it was something else that glimmered in their depths that put a finger on my heart. I grinned. He wagged his tail. I leaned down to pick up an oversize paw. Exploring fingers ran over the coarse pads. They seemed as rugged as they were big.

"Well, he's all yours," affirmed Mr. Corson, as he unchained the dog. "His name is Black. I don't know what kind of a hound he'll make. I let him loose a coup'la times and he chased a neighbor's cat. If he's no good, you can shoot him." He did not speak callously, only matter-of-factly. "Come into the house. I'd like your name and address, and I can't read nor write. My wife does it for me." Neither apology nor embarrassment toned his voice. "Besides," he continued, writing his autobiography with a single sentence, "I'd like to give Black a bite to eat before he goes."

Warming my hands around a near-boiling cup of coffee, I pondered the trek back to my truck. Five drifted miles separated us. Here was a trail hound who had never seen me before. Suppose he took off after the first track we came across. What would I do then? It was late afternoon and I'd rather spend the night in bed than in a snowdrift. I turned to my host.

"Just in case Black doesn't want to stay with me, I think I'd be better off with a leash. Do you have an old piece of rope I could use?"

He didn't answer me right away. He looked from me to the dog and back again. He smiled. "That won't be necessary," he predicted. "He'll go with you. You won't have any trouble."

I didn't really believe him, but since he insisted, I spoke to the dog, "Okay, Black Guy, let's go home!"

The hound followed me out the door. He wanted to come with me. It was almost as though he had been waiting for me. He did not touch me. He did not speak. But, while we tramped back to the civilization of plowed roads, he somehow seemed to become my dog. When we reached the vehicle, I opened a door, and he jumped in. Most dogs jump onto a car seat, but when I told Black to sit on the floor, he obeyed. During the ride home, he did not take his eyes off me. I had never met a dog like this one. I recalled Mr. Corson's discerning smile. As a dog man, he must have sensed that something special had happened that day in the old barn.

A hound dog needs a short, mouth-filling name. Because bounty hunting can often be dangerous, there is seldom time to shout but one terse command. Besides, you won't get hoarse as quickly while hollering for an obstinate, "I-know-more-than-you-do" partner. Training Skivvie had taught me that. The new arrival's name filled the vocal bill, but a man needs to name his own dog. This time it seemed especially important. Ben, Tom, Reb, Pete, every name I'd ever heard, rolled over my tongue, only to be spit out. A week passed.

"Well, Black Guy," I puzzled, "I guess the jig is up. Looks like you're going to stay Black. I can't think of a name to save my soul." Suddenly, I laughed aloud. That was it. I'd call him Jiggs.

I expected him to be confused, but after a first quizzical look, he wagged his tail and galloped over to bump against me.

The family adjusted quickly to its newest member, even the hard-headed Skivvie. After the usual stiff-legged, hackles-up examination that is part of being a male, Jiggs accepted the older dog's seniority without question. As a traditional southerner, Jiggs had perfect manners. Friendly, but never fawning, he demanded instant respect, with affection naturally following soon after. My daughters had been taught that dogs, like people, are individuals whose dignity must be respected. Because, as children, the girls were about the same age as he was as a dog, the three had much in common. Within days, they became fast friends; but affectionate as he was with them, he made it clear from the beginning that he was my dog.

He wore a new, handmade leather collar with an oversize brass nameplate riveted onto it, upon which his master's name, address, and telephone number were etched. Big, bold letters are essential. Tags on dog collars are worthless; lost hunting dogs invariably come into the last farm on a dead-end road, and farmer's glasses are often spattered with either milk or manure.

Before letting Jiggs begin his apprenticeship with Skivvie, I decided to see what "junior" had to offer at the ten-months level. At daybreak, we were up and headed for a chopping on the west face of Willard Mountain. As we began to follow an old, snowed-in tote road, he discovered how touchy I am about sharing my snowshoes. Rarely out of my sight, he cast continuously around me. Suddenly he stopped, faced downhill, and lifted his muzzle into the air. A light west wind had blown the odor of something delightful up the mountain, and he started to cast in the direction his nose took him. Because of his complete innocence, I watched with amused anticipation until he disappeared into a heavy stand of young conifers. The attraction was obvious: several deer always yarded up on this side of the mountain. I waited a few minutes, then resumed my climb. The road corkscrewed around a big chopping before reaching a plateau. There I stopped to wait, uneasy because a half-hour was too long for an apprentice to go unprompted, and because a deer yard was not the place for him to learn how to ad-lib. Then, from my seat in the balcony, I watched the drama unfold.

Out of the heavy spruces, back into the tote road, dashed Jiggs. He looked up, then down. Where had his meal ticket gone? Long ears flying, he sprinted back and forth at breakneck speed. Because he could not see his master in the flesh, he panicked. Too young to consult his nose, he ran pell-mell, this way and that, hoping to bump into him. After ten minutes of complete confusion, with me grinning up in my box seat, he stopped, moved back down to the road. For the first time he put down his nose. He flew up my snowshoe tracks like a startled grouse, but when the road turned, he did not. Again he went berserk while searching for my body instead of my scent. He sprinted back into the open, found my smell, and dashed up the trail. He made the same mistake on the next bend, came back out almost at once, searched out my smell for the third and last time, and galloped up the last few hundred yards as though on the home stretch in the Belmont Stakes. When he saw me standing by a pine tree, waiting, he bounded over, wagging his tail delightedly. He had just discovered a new world, a world waiting to be explored with his nose as the compass, his brain, the rudder. It was also a day of discovery for me. From the first moment, my voice was his leash; so from now on, he ran free. No other dog ever earned this privilege.

A few days later, he learned his second lesson on the slopes of Bald Mountain. Although the track we found smelled of cat, a great deal could not be expected. Jiggs was too young, but what he lacked in years, he made up in enthusiasm. He wanted to try. Excitement charged my command of "Get him!" He bolted. Within moments, he began to bay and disappeared over the mountain. "Too good, too fast," I said to myself as I clambered up to check. To my dismay, he had followed Big Kitty's trail into a deer yard, jumped the tenants, and because a half-dozen deer use more perfume than a lone cat, the choice had been easy. Fortunately, the deep snow became my ally, for the deer made a loop, then headed back for their yard. A dead sapling was hastily broken off. I reached the yard just in time to have seven deer almost trample me. A minute behind bawled my junior partner. A flying tackle and a mountain-shaking oath were followed by the only thrashing I ever gave him. One of the most important lessons any hound dog must be taught is that, for him, deer are forever sacred cows. Jiggs cowered at my feet and took his beating. He did not try to run away. When I finished, he was a subdued dog.

"Okay, young fella'," I encouraged. "Let's go home. You've learned enough for one day."

Lesson number three put theory into practice. We had been hunting for several hours, a tractable Jiggs using his nose continuously as he cast about. Across a hardwood clearing I noticed tracks. We were walking through a deer's front parlor. I broke off a switch, waited until my pupil got in ahead of me, then continued behind him. He pushed his muzzle into the old deer track. Down came the switch. He yelped, jumped behind me, and looked up as I admonished sternly, "Not supposed to, Bad Guy." Looking to neither right nor left, he followed me across the deer's trail. A half-hour later, fresh tracks, reeking with scent, tested him. Ahead of me, Jiggs suddenly dropped his tail, slipped behind me, and remained there until we were well beyond the forbidden fruit. This is all it took to teach him deer were off limits. During the ensuing nine years, many, many times he coldtrailed cats through the middle of a yard harboring more than a dozen milling deer. They might just as well have been a neighbor's cows.

The first time the two dogs hunted together, Skivvie proved he was as proud as I already knew he was stubborn. Pressed by a merciless subordinate, he gave the quarry as little quarter as Jiggs was giving him. Within an hour, the blast of my shotgun ended the chain reaction. A week later, the same stubborn pride resulted in tragedy.

Both dogs had bulldog tenacity, but where Jiggs persevered, Skivvie persisted. He had pride. Jiggs had self-respect. The inevitable contest lasted five days and four nights while they hunted within a thirty-five-square-mile triangle. On the first day, when they had not returned to the truck by dusk, I was worried. To keep them away from the highway, I put down a blanket smelling of me and of home. They never used it. On the second day, an ice storm armor plated the area. For three days, I rode every boundary, asked every farmer, hiked every tote road, and hollered from every mountaintop. On the fourth day, an old-timer, who lived in the isolated area, raised my flagging hopes.

"Come to think of it, John, two days ago a coupla' fella's from the village were hunting porcupines six, seven miles down the pond. Said they heard two dogs drivin'. Figured sure they were chasin' deer, so they went in to look around. The hounds were runnin' a bobcat."

I hiked down along the lake. The three-day-old tracks were theirs. The smaller ones showed red. Much of myself had gone into training Skivvie. As for Jiggs, quite by accident, I had acquired a once-in-a-lifetime dog. Now I had lost him—maybe for good.

Around noon on the fifth day, the phone rang. A voice twanged, "Mr. Kulish? Say, I got your dog here. No, just one. He's black. I drive school bus up here. This morning, when I stopped to pick up the kids on Route 9, a big, black hound trotted out from the shed and got on the bus with the kids. He jumped right up on the seat and sat down as friendly as you please. The kids wanted to take him to school, but I brung him home and fed him. He sure was hungry." He paused, "Say, you wouldn't want to sell him, would'ja?"

When Jiggs saw me, he stood up. Tall, skinny, tired, he wagged his tail and bumped his muzzle against my leg. He never did get giddy about anything.

The next month was torture. No one had seen Skivvie. I scoured the area, but he had left no traces. He must have followed a cat into a ledge and not been able to get back out. I blamed myself. I should have realized that, when challenged, Skivvie's tough heart would outrun his tender feet. I slept badly. I ate little. The veterinarian told me it might take a dog thirty days to die of starvation.

9. He Earns His Ph.D.

By the time he was three years old, Jiggs had several encounters in combat severe enough to take him to a veterinary. Young Dr. Donovan, who had recently moved into town, tended all his wounds. My partner always knew when he was going to the "vet's." "Well, Kitty Guy," I'd drawl, dragging out each syllable in the special Black-and-Tan dialect I used when talking to him, "it looks like you are going to see Dr. Donovan again."

Several skirmishes with wounded bobcats had ended on the doctor's operating table. One time an ear, punctured and torn by a dying cat, had to be stitched. Another time, after a cat had ripped off a big chunk from the other ear, Jiggs had to spend seven days in the hospital under sedation. Once a day, he was roused only enough to eat. Jiggs hated going there. Separated from me, confined in a cage, and surrounded by cat smell, he was miserable. But he never rebelled, and the "vet" said he was the easiest dog he'd ever attended. Each time he had received a Purple Heart, it had been preceded by a trip to Dr. Donovan's.

Sooner or later, all hounds, especially those trained to hunt cats, have a face-to-face encounter with "Prickly Porky." Jiggs was three years old before his first joust with Sir Porcupine. They crossed swords on Round Mountain, six snowshoe miles through cat country from a plowed road.

Just before noon, Jiggs entered a blowdown above the glacial rocks buttressing the base of the mountain. A hundred yards behind, I struggled up, over, and under the tangle of dead trees. I had stopped for a breather, when a growl, followed by grunts, snarls, and thrashing noises startled me. What had my partner encountered? I clambered up to reach him. Pieces of porcupine, quills, and blood were strewn over the snow. His head down, Jiggs swiped desperately at what looked like ten thousand Tom Thumb spears that had penetrated his muzzle, jaws, and neck.

I grabbed the writhing dog by the collar and forced up his head to look him full in the face. My heart tightened. Pain-filled eyes pleaded from a grisly pin cushion. As porcupines grow older, their quills grow longer, and Jiggs had not assaulted a baby. These spines were four inches long. Junior packs a .22; Grandpa carries a broadside of 30-30's; Kitty Guy had the full charge at point-blank range.

He hurt so I could hardly bear to look at him. Even his paws looked like chestnut burs. For a terrible moment I raised my shotgun, but could not pull the trigger. Could I get him to Dr. Donovan before it was too late? In his anguish, Jiggs paid no attention to my commands; so I snapped a leash onto him. He still wouldn't budge, but lay on the snow and clawed at his grotesque head. He moved only when the collar pulled so tight that it was either walk or choke. Spurred by fear, I tugged, dragged, pushed, lifted, and carried him the longest six miles of my life. When we finally reached the road, the truck was barely discernible in the gloom. The twenty-five miles to Dr. Donovan's seemed like a hundred.

It was long after hours, but since the doctor's office was attached to his home, he told me to carry the dog into the operating room. The shocked look on the veterinarian's face undermined what little hope I had. He said he'd do all he could. I held Jiggs in my arms while anesthetic was injected, then gently lowered him onto the table. The doctor opened wide the flaccid mouth. The tongue was nailed to the floor and roof of his mouth by countless quills, some of them broken off. They pierced his jowls and nostrils. Battalions of them marched out of sight down his gullet. The doctor called his wife in to help, then promised to call me as soon as he could.

Hours later, the telephone rang.

"John, Jiggs is alive, although I don't know why. It took the two of us three hours to pull out the quills. There must have been more than a thousand. I'll never understand why he wasn't blinded." Then, with a smile in his voice, he added, "You can come get him in the morning."

Each day, as Kitty Guy's mouth and tongue began to heal, my searching hand would pass over the swollen muzzle, feeling for any broken quills that might have worked their way to the surface. When three days had passed without the emergence of a quill, my breathing became easier. Each morning, when I started to lace my boots, Jiggs would bang his tail on the floor, telling me he was ready to go back to work.

A week after his near-fatal ordeal, we were back in the woods. I was snowshoeing along the backbone of a minor ridge. Fifty yards below and paralleling it, was a narrow ravine where Jiggs was zigzagging in search of a track. My heart lurched. Seventy yards ahead, a huge porcupine was waddling across the ravine at right angles to the dog. Jiggs froze. His ears moved forward into battle position. He could see a creature moving but because a slight, downwind breeze was blowing, he could not smell its identity.

Once again I could see him on the operating table. Once again I could feel the agony. Only an instant remained before the attack. Both hands cupped my mouth. The silence was shattered by a shout, "DOC-TOR DON-O-VAN!"

Jiggs's ears uncocked. His head dropped. Tail between his legs, he backed up slowly, looked up at me, then slunk up the knoll to get into my snowshoe tracks. Whoever the enemy, he wasn't worth another Purple Heart.

In battle, Jiggs was a Sir Winston Churchill. The greater the odds, the fiercer the fight. Nor was he afraid to abandon a lost cause—temporarily. He had had a few Dunkirks of his own. They just made him hunt harder. This is where brains took over, for Jiggs could make decisions and follow them through. Each day, I knew what I could do, and I knew my partner would do even more. Because we were both willing to go until dark without stopping to eat or to rest, cats didn't get much sleep.

Whenever tracks were scarce, the whole load shifted onto his rugged shoulders. After as many as six barren days, we'd stare down at snowed-in spoor. Jiggs would study me while I studied the evidence. Big Kitty was many hours and miles ahead, but suppose a fresher track crossed this one up ahead? Should we gamble? If we followed the trail together, I would set the pace. But in cat country, four-wheel drive is better than two.

"Go get him, Kitty Guy," I'd decide. "Shift into high. I'll back you up."

One could not do this with an ordinary dog, but I trusted him completely. By sight, he could tell the difference between a cat track and one made by any other creature. Even the gray fox couldn't fool him. If, after a half-day, our gamble paid off, he would take the better track without hesitating. He made decisions on his own exactly as though I were there, using my own knowledge, based on my own experience.

A ten-mile scouting trip into enemy territory was not unusual for him. He gave up only when he found his quarry already bedded down in a ledge, or when it had crossed a frozen lake that the wind had swept clean. Terrain and weather made no difference to us. We just modified our tactics.

Some winters were worse than others, for Mother Nature makes many different kinds of snow. The heavy, wet variety that packs as it falls, is a snowshoer's delight. Thaw brings crust. The breakable kind shatters like glass, drawing blood from men, dogs, and snowshoes. Occasionally, we got buried with powder. For weeks on end, zero temperatures revived only long enough to dump another foot or two of feathers onto a base that was already impossible to hunt in. It was like wading hip-deep through down. Jiggs would plunge in and swim, leaving a trail as though a prime log had been dragged through, except for the tail marks that flopped from side to side. Finding a track was impossible. Cats had as much trouble snowshoeing as I did. But, day after day, we would persevere. I had an A-1 dog, legs like a mountain goat, and a mortgage on my house.

One time, it took us three days to pack down a three-mile-long trail into a rabbit swamp. Each morning we took up our highway construction where we had left off the night before.

"Stay in my trail, Kitty Guy," I urged. "Save your strength for what's ahead." But, giant that he was, he chose to shack on either side, his rime-covered muzzle a periscope in the sea of snow.

Late on the third day, we plowed into snowshoe hare country, where rabbits were born, lived their two-year life spans, and died without ever seeing a human being. This was also the hub of bobcat universe. In the tangled alders between spruce-covered knolls, the hare had beaten down troughs a foot deep along their feeding trails. The day before, a cat, walking gingerly in them, had sunk but an inch. I could almost feel cat close by. Jiggs began to bay. Back and forth he crisscrossed the supermarket, with me weaving about in the aisles trying to intercept the customer. Before long, both dog and cat tracks were appearing where I had been only minutes before.

"Well, John," I thought, "it's time to pick a checkout counter." I climbed a likely spruce knoll and listened as Jiggs drove closer and closer. Suddenly a cat, using the trail I had just made, darted toward me. The blast almost singed Jiggs's whiskers. Our three days of road building had paid off at $6.67 a day.

During the fourth year of our partnership we hit our stride. One morning we were up, had driven fifteen miles, snowshoed three, and had jumped our first cat, by the time most other people were thinking about having breakfast. High up on Osgood Mountain, the pulse-quickening sound of Kitty Guy's "I can see him, John" bark, urged the safety off my gun. And not a moment too soon, for a twenty-five-pound bounty leaped up through the rubble toward me. The gun spoke, once, then twice, and both times with authority. The cat disappeared behind a boulder. Turning to welcome Jiggs, my eyes popped. A second cat, as big as the first, dashed toward me. One shot dropped it, and before another round could be pumped, Jiggs appeared. To my amazement, he ran by the still quivering cat and out of sight behind the boulder where the first one had disappeared. I scrambled up the rock pile. Jiggs and the cat were locked in a death struggle. The dying cat had wrapped its claws around the dog's head. As they thrashed about, snarls and fangs intermingled. It was too risky for me to shoot. No club was available so, raising the stock of my shotgun, I waited my chance. The weapon crashed down on the cat's skull. Spattering blood, Jiggs shook himself free. Except for an ear torn in the struggle, he was intact.

"Kitty Guy," I apologized, "next time, I'll either shoot straighter or run faster."

Full-grown cats don't usually run tandem. Unknown to each other, these two had been hunting the same swamp at the foot of the mountain. Soon after Jiggs had jumped one, it headed for that rock pile. When the chase threatened the second cat, I discovered that scared cats run in the same channels. Because cats have such poor noses, they couldn't smell each other. Barely out of one another's sight, they sought safety in the same sanctuary.

To a hound, each kitty has its own special perfume. As far as Jiggs was concerned, there was but one cat to chase—the one wearing Chanel Number 5. The second cat surprised him as much as it did me.

By one o'clock the two bounties were cached in the truck. Blood still dripped from Jiggs's ear.

"Well, Kitty Guy," I suggested, "how about tackling their third cousin up on Thunder Mountain?"

When we caught up with the relative, visible flames shot out of the barrel of my shotgun into the night. As a crow flies, I was nine miles from Osgood Mountain. I wasn't a crow, but a man in a hurry with a pack, a gun, and a third dead wildcat. Above me, the tip of the mountain punched a hole into the clouds; snow spilled out. Now came the chore of finding my way back through the dark to my truck, four miles away. Human eyes that are used to the out-of-doors develop antennae. A couple of hours later, with the help of my compass and a reassuring muzzle thrust regularly into my mittened hand, I reached the highway.

My wife ran out of the house to greet me. "I'll bet you don't even realize it's nine below. As a matter of fact, it never got above zero all day; the wind blew so hard I had to fight to open the shed door." Her eyes widened when she saw a month's groceries stretched out in the pickup. She hugged Jiggs. "John," she chuckled, "when it comes to picking women and dogs, you're in a class by yourself."

After supper, as late as it was, I called the local conservation officer. State law requires that dead cats be turned over to a game warden within forty-eight hours. I wanted a few hours' grace. Three inches of powder had already erased today's writing. It was still snowing. I wanted to start tomorrow's story early.

The next morning I surprised the sun above timberline as it popped over Mt. Monadnock, making it blush so hard the snow fields turned rosy pink. Before noon a thirty-pound tom was stretched out behind the truck's seat. Then, after a thirty-mile drive north, Jiggs and I raced the sun to reach an isolated swamp offering fresh rabbit à la carte. Two hours after dark our fifth cat in two days joined a relative in the pickup.

The next day the conservation officer arrived. I helped him carry 135 pounds of cats from my garage into his car. Jiggs and I had stretched our sinews that weekend.

Kitty Guy's fame spread and, when he was seven, I refused a thousand dollars for him. "Would you sell your wife or your children?" I asked the would-be buyer. My wife summed it up cheerfully, "If Jiggs could cook, I'd be out of a job."

For me, one trademark of a true cat hunter is the improvised rawhide sling on his shotgun. Inching up over icy rock piles or hauling oneself up mountainsides so steep the butts of the trees one clutches are almost horizontal, requires both hands. So does grabbing tree trunks to break one's plunge back down over the ledges.

That morning, I thought wryly of my clients who gush, "Boy, John, are you lucky to earn your living the way you do!" Ten inches of fresh snow covered the frozen ground—not quite enough to use snowshoes comfortably; a little too much for walking without them. In the half-light I looked up at the shadowy string of mountains looming north and south four miles ahead. Hopefully, a cat track waited near one of the half-dozen peaks. A couple of hours, two swamps, and three blowdowns later, I started toward the fourth blowdown, which, in previous years, had often produced the spoor I was looking for. I had picked my way up a quarter-mile of cliffs when Jiggs looked down at me from a ledge. Cocking his ears and stiffening his muscular body was his way of telling me he'd found a good track. By grabbing the stunted trees growing aslant out of the steep slope, I hoisted myself up to his level. He had indeed found a cat track, one barely an hour old.

"Get him!" I exhorted.

Within minutes, he began to bay in the blowdown above. Up and down the mountain they went, traversing the base four times before racing out of hearing. Soon the baying came back, to veer away once more. I moved further among the windrows of fallen trees which crisscrossed the slope. Through the latticework beneath me, I could see a mosaic of cat and dog tracks. This was the place to grab the brass ring on this merry-go-round.

Suddenly, not more than twenty yards away, the cat jumped noiselessly onto a fallen tree, head turned in the direction of its pursuer. The moment I moved, it would see me. My hasty shot blew the cat off the log, just as Jiggs arrived underneath the blowdown to claim his prize. I expected to hear the thrashing sounds of him shaking the cat. Instead, he yowled and plunged down the steepest side of the mountain. I hop-skipped-and-jumped to where the cat had somersaulted off the log. The snow was spattered with blood. The victim had dragged itself down toward some glacial rocks, piled boulder upon boulder, two hundred yards below. Jiggs had stopped baying.

At the upper end of the half-mile-long rock slide a frantic dog met me. He was running round and round the bloody trail where his quarry had dragged itself under the foremost boulder. I could hear the cat's labored breathing, then the death rattle.

Now only silence. To me, this wasn't just a cat; it was my day's wage. I had a wife, two children and payments to meet on my house. I knew Tom wasn't far underneath, but the opening was too small, the rocks covering it too big. For more than an hour I tried in every primitive way available to get at my day's pay. With my jackknife I cut down a long green sapling, trimmed off the branches except for a single sharp hook on one end, and pushed this into the crevice under the boulder, hoping to snag the cat. It had been done on other rock piles under similar circumstances. The sapling bent underneath the boulder. I located the obstructing stone, but because it held a much larger one on its shoulders, there was no chance of getting at the bounty. If this one stone could be removed, I could hook onto Tom. I finally gave up—but only for that day. I didn't intend to throw in the towel after the first round.

It was midafternoon of a January day and I had left my pay in a rock pile. How to get it? During the summer months, road agents in our small New Hampshire towns have a use for dynamite. In winter, whatever is left over is a nuisance because, once frozen, it loses its effectiveness. My own town was larger than the average village and had a correspondingly well-equipped highway department. It must have a surplus of dynamite. I understood explosives, because, during summers, I had done blasting for a living. My spirits soared. I knew what to do. But Jiggs was disappointed. He had failed to sink his fangs into the warm body of a bobcat—a big one at that.

"Good Kitty Guy," I comforted, patting his canny black head. "Not today. Maybe tomorrow."

At dusk I knocked on the road agent's door. Henry had twenty-two sticks of dynamite. He was delighted to have me take them off his hands for a dollar. That left nineteen dollars in the rock pile and furnished me with a big surplus of power. After supper I visited a friend who has a complete workshop. Alfred was as interested in tools as I was in animals. With his know-how and my specifications we produced a rod about six feet long and a quarter-inch in diameter. Holding it in a vise, he fashioned a pointed hook on one end and then twisted the other into a looped handle. Now I was fully prepared.

The next morning, it was still dark when I waited while Jiggs went through his before-the-hunt, tree-to-tree ritual. Two sweaty hours later we reached the rock pile. I took the dynamite out of my pack, wrapped a cord around the sticks to form a solid bundle, took out the fuse, inserted it into the blasting cap, punched a hole into one of the sticks, pushed the cap in, and folded the dynamite gelatin over the rupture. With a long pole the package was pushed under the blocking boulder. Jiggs had been eyeing me all the while. I snapped a leather leash onto his collar and hooked the loop onto a sapling to make sure he would stay with me. This was no time to take chances.

The twelve-inch fuse was lighted. It started to sparkle. I scooped up my gun and pack, grabbed the Ol' Guy's leash, and sprinted three hundred feet. There, parallel to the rock pile, I crouched behind a two-foot-thick beech tree, making sure that Jiggs was firmly between my legs.

A roar shook the ground. Even the beech tree trembled. Waves of sound banged around the mountainsides. Large and small stones rained in an angry shower. Jiggs delighted in the blast of a shotgun. But he'd never heard a noise like this one! Released, he raced toward the disrupted rock pile. A halo of blue smoke hung over the hole. Rubble lay scattered everywhere. The big boulder had splintered and rolled away, opening the cat's tomb. I got down on hands and knees to push aside the broken stones. It was still not possible to see our quarry. The iron rod was guided into the opening, and moved around at arm's length until it touched something which wasn't stone. Now came the job of hooking onto the body and dragging it toward me. It slipped off. I lay prone to reach with my arm up to the armpit. Fingers touched the hind leg of a twenty-dollar bill, minus that dollar for dynamite.

10. *The Longest Day*

Most hunters envy a guide's ruddy complexion, covet his flat belly, and delight in his earthy, backwoods philosophy until an easy shot is missed or an animal wounded. Then they learn that the woods hath no wrath like a guide listening to excuses. Guiding taught me much about human behavior, especially my own. I'm still not sure whether I learned to understand people because they are so much like wild animals, or wild animals because they are so much like people.

116

All the men I guided shared a common emotion about my way of life, even though some of them earned more in one month than I did all year.

"Boy, John, do we envy you! You don't know what it's like to be frustrated. How can you understand days in an office when nothing goes right?"

Move over, gentlemen. There are days when a professional woodsman also wishes he had never gotten out of bed. Consider that day on Surry Mountain.

It had started out like any other day. As I struggled to strap on my snowshoes, there wasn't enough light to see the buckle holes in the harnesses. Shotgun loaded and slung over my back, the three of us started up the only gradual approach to the summit. There were three, for this morning, Kitty Guy's son, Chocolate, accompanied us. Though only sixteen months old, he had all the makings of his sire. Father was going to teach son. There's no better way—if Ol' Dad knows his stuff.

The presence of a younger partner roused a typical male reaction in the older dog. Casting on either side of me, the dogs ranged farther than usual. The morning was middle-aged when they "opened up" on a track so hot they had vanished out of hearing by the time I reached the jumping-off place. In the snow were the tracks of a pair of rutting cats. A quarter of a mile up the mountain, the two separated when survival had overcome sex. Both hounds stuck to one kitty. I looked at my watch. It was nine-thirty. My chances of bagging cat number one were good. Once treed, there would still be time to retrace my trail to try for the other. Confident of a forty-dollar day, I took off, whistling under my breath and with an extra spring in my stride.

For the next two hours, the hunt took me over ridges, down gullies, through swamps, over the crest into a valley, back over my own trail, and finally, to the brink of big Surry Reservoir. A grin wrinkled my cheeks as a west wind blew the sweet sound of hounds baying "treed" toward me. It was midday and the first bounty was up a tree. Eagerly, the descent began, but because the ridge was so steep, my snowshoes became a hazard. Long, narrow, with turned-up toes, they responded to the pitch like skis. Off they came, to be upended in a snowdrift.

I slid down the precipice to reach a narrow shelf running across the face of the slope. Moving cautiously along ice-encrusted rock, I reached the dogs. Bawling and bounding, they were trying to climb a tall, straight pine tree. From forty feet above, a large cat's tiger eyes glared down. Because the first shot is the one that counts, careful aim was taken. I squeezed the trigger. A flat click startled me more than the accustomed roar. Berating the inspectors in a shell factory, I worked the pump, took deliberate aim and pulled. Click. Frantically, I pumped the action and yanked the trigger. Two more shells flew beside the first three. Groping in the snow, I picked up two of them. No impression from a firing pin indented the primers. The breech flew open; the bolt was examined. The firing pin was gone!

What should I do? Where could I get another gun? How long would it take? Would the dogs remain at the tree after I left? If Chocolate, young and impatient, left, would Jiggs stay behind? But leave the tree I must. Never had the bitter truth of "so near and yet so far" been more clearly understood. A rawhide thong was pulled from my game pocket and fastened around Junior's collar. My only hope to use him as staying power was to tie him to a nearby tree. But how long would any dog remain under these circumstances? Jiggs should stay for some time, for he had learned from previous experience but, veteran that he was, here was an area that remained untested. The useless gun was hung onto one of the lower branches of Tom's roost. If both dogs deserted, the cat might still be unwilling to come down past a gun reeking of man. This was a wild gamble; bobcats are notorious for their poor "smellers."

Where could a weapon be found? No one in the vicinity knew me. Then a thought cheered me. The reservoir was formed by a federal flood control dam. A permanent dam tender is always stationed at such installations. Perhaps he would have a gun that

I could borrow. The dam waited three miles below. My snow-shoes waited a quarter of a mile above, where I had so optimistically left them. Climbing up the cliff would eat up precious time. Suddenly, looking down at the reservoir, it didn't look that far away. Mirages never do.

Although two feet of snow had covered the icy base formed when winter's first snows had melted, it was but a camouflage for the treachery below. Because of the sheer pitch, some cliffs could only be traversed. Sliding down others, I would gather up momentum and crash violently into a tree. After several bruising collisions, I began to pick out certain trees that loomed closer, and by plunging from one to another, I ricocheted down the mountain.

I had but a few moments to relish the thought that only a quarter-mile of level ground separated me from the reservoir, when Chocolate, pink tongue lolling, wig-wagged down to me. Was mine to be a fool's errand?

Wallowing in thigh-deep snow, I thought that the pond would never be reached. The windswept mantle on the reservoir was just as deep. It could not hold me. As soon as all my weight was placed on one foot, I would sink almost to my crotch. The gate house wavered miles away. Stopping to rest, the weak baying of a hound could be heard far up on the mountain. It seemed like hours before I made my way with a jerky, muscle-tearing stride up the riprap to the top of the dam. The plowed road brought tears to my eyes.

I flew over the half-mile to the gate house. In the maintenance building, a man was working. He turned out to be the assistant dam tender and, as such, did not live on the government premises. My predicament was explained. He did not seem sympathetic, even when interjecting that his superior had a shotgun, although he didn't know what kind. In typical Yankee fashion, it was only when I said I would hurry along to the dam tender's house that he told me his boss was out of town and would not return until late that night. Would he get me the gun? No? He didn't want to go into his boss's house? Disheartened, I started down the road. Surely, someone would help me.

A half-mile below the dam, I knocked on a front door. A dog barked from a shed. I knocked on the side door. I knocked on the back door. Wasn't anyone in Cheshire County home? A desperate half-mile farther brought me to the next house. Judging by winter sign, it didn't seem to be used a great deal, but a car was parked in front. A middle-aged woman answered my knock. My troubles were dumped onto her solid shoulders. It took the better part of five minutes. After I had finished, she told me that the owners were down South. She checked the house once a week. No, she didn't know anything about guns and cared less. Refusing to consider myself the victim of a plot, I continued down the road. Chocolate jogged along beside me. No one answered at the next door. The knowledge that I was going farther and farther away from my cache on the mountain maddened me. The empty highway stretched ahead.

Suddenly, a vehicle approached from behind. It was the assistant dam tender, driving a government truck. He had decided to chance his supervisor's ire. While I waited, he unlocked a door and went into the house. Ten minutes later, he reappeared. He held a rusty, double-barreled, twelve-gauge, hammer shotgun. He had no shells. He had looked everywhere. Driving into Keene with a government truck was out of the question. One man in trouble was enough. Then he recalled once seeing a shell or two lying around in the maintenance building, so we drove there. Again I waited. It seemed like a long time. When he came out, he held up four shells. I grinned. It wouldn't take four shots to shoot a cat out of a tree.

A second glance revealed that two of the shells were useless. They were No. 8's, dud shot for killing sparrows. The other two were 00 Bucks, a coarse pattern used for deer. The nine individual balls have such a dispersion that one can miss a moose at fifteen yards. Compelled to become a reluctant chooser, this beggar took the 00 Bucks. Putting the ancient blunderbuss under my arm, I hurried across the dam.

Climbing a steep mountain without snow is difficult. Climbing a steep mountain with snowshoes is worse. Climbing a steep mountain in deep snow without snowshoes is a horrible experience. My best bet was to intercept my own snowshoe trail somewhere on the east slope, follow it to the dropoff, and thence down to the cat. After an enervating struggle, the summit was mine. No barking could be heard. Dejected, I slipped and slid down past my snowshoes still stuck in the snow, to reach the rock shelf. During the cautious approach toward the pine, I suddenly heard a hoarse bark. Ol' Kitty Guy had passed another test.

My broken shotgun still hung over a broken branch. A twenty-dollar bill still glared down at the hound. Jiggs had done his job; the rest was up to me. I opened up the breech of the old double-barrel, loaded both chambers, and snapped it shut. What if I fired one shot and missed? What if only a single pellet grazed or struck the cat? Past experience had taught that such a cat would leap out of a tree, no matter how high, and disappear down the cliff below into any one of a hundred rock piles. The most important decision in a day full of decisions was made. Eighteen pellets are twice as many as nine. Both triggers must be pulled at the same time.

A steadying boulder was essential. I did not trust an offhand shot. The hand supporting the barrel resting on a rock, both hammers at full cock, I took careful aim. A roar shook the mountain. The cat spiraled through the air end over end, and disappeared into a coniferous thicket below. Dashing to the foot of the pine, and dropping to my knees, I peered down the cliff to where my day's pay had disappeared. The mountain was so steep that when the cat hit the snow seventy-five yards below, it slipped out of sight like a toboggan. Jiggs leaped after his trophy, be it dazed, wounded, or dead. The empty weapon was leaned against a tree. Lying on my back, my spread-eagled arms and legs as brakes, I slid down 300 feet.

I had a cat all right, but even as Jiggs extracted payment, the problem of dragging it back up prevailed, for Chocolate had broken the leather thong used for that contingency. A rawhide lace was removed from one of my hunting pacs, a loop formed, and a noose fastened snugly around Tom's neck. Crawling, reaching, slipping, grabbing, it seemed as if the pine where two shotguns waited was beyond my capacity. A desperate heave lunged me up onto the narrow ledge. I lay on my belly, fighting for breath. The two weapons were added to my load and the final assault begun.

Both hands busy with bulky burdens, I had to resort to creeping, sometimes almost losing a shotgun. My heart pounded whenever one began slipping out of my hands. I didn't know which gun to throw away. Somehow or other, my snowshoes were reached. Encumbered with one thirty-five-pound bobcat, two five-foot-long snowshoes, two ten-pound shotguns, two large dogs, and one unlaced, snow-filled boot, I finally conquered the pinnacle. Under the first stars, I flopped down onto the snow to rest and to think. What had promised to be a beautiful, forty-dollar day had turned into a tortured nightmare. I felt like a man who had been out hunting for the first time.

As an imperfect perfectionist, I hunt alone, taking pride in the orderly sign left by snowshoes that, forming a giant loop, make but one tidy, continuous trail. That day, the whole mountain was covered with tracks: all made by one man. What was it that Henry David Thoreau had said about "most men leading lives of quiet desperation"? I wish he could have been with me that day on Surry Mountain.

11. Give Your Heart
to a Dog

The relationship between Jiggs and myself deepened. With every passing day, we understood each other better. I was more than just a man to him. He was more than just a dog to me. During the fifth year of our partnership, high up on the ice-covered cliffs of Mt. Monadnock, I discovered how much the Ol' Guy had come to mean to me.

As a woodsman, I frequently faced physical danger. When face to face, the fear I felt was usually sudden and short-lived. It demanded instant action, while adrenalin applied mouth-to-mouth resuscitation to ebbing courage. But that dark day on the mountain, fear of the unknown, fed by uncertainty, made my very heart buckle.

Each day I hunted a different township, visiting a certain area but once a week. Mt. Monadnock was on my regular schedule. In late autumn, bobcats had hunted the foothills of the 3,200-foot peak, but during deer season, they had withdrawn to timberline.

Before daybreak, Jiggs and I left a dead-end road that winds partway up the mountain, and started a two-mile climb toward the ice-encrusted, bald summit that dominates the region. Four inches of fluffy, new snow covered two feet of packed powder. Rabbits and deer had taken advantage of perfect feeding conditions during the night, but cats had not. Jiggs left me regularly for ten- or fifteen-minute intervals as he made consecutive circles. Two hours later, we had climbed within a quarter-mile of tree line. As I scaled the ledges on the steep east slope, I murmured a thank-you to the small, tough saplings that held firm, though supporting my full weight.

Jiggs had been gone for a half-hour. Perhaps he had found a track made early last night and was trying to get Tom up. For almost an hour I waited restlessly, all the while studying the weirdly beautiful snow formations the winter gales had sculptured along the ledges. The frozen sun and my grumbling stomach told me it was noon. Still no Jiggs. Had he jumped his quarry and taken it over the mountain out of hearing? To keep from freezing, I continued to climb slowly toward the summit until, an hour and a half later, I stood on top of Pumpelly Ridge. Unmarked, the snowy rock fields undulated north, west, and northeast. Puzzled, I snowshoed north along the slippery crest for another hour, searching either side for a clue, stopping regularly to listen. Never before had I stood here without fighting the wind. It had retreated completely. Not a puff stirred the snow. The unnatural quiet disquieted me. Pacing back and forth, I watched and listened atop the northernmost peak of the ridge until the numbing cold drove me back.

Gathering clouds obscured the setting sun. I could wait no longer. Before dark, I must try to reach the place where Jiggs had left me. I began a cautious descent; haste could result in disaster. Thoughts kept my mind as busy as the tricky ledges kept my legs.

Jiggs had been my partner and constant companion for five years. I knew him inside out. Never before had he behaved like this. Could he have jumped a cat on the south slopes of the summit and run it off toward Gap Mountain? But why had he broken precedent to range so far way from me?

It was almost dark when I reached our parting place. Only this morning's tracks marred the snow. It was too late to search further. Descending the cliffs at night would be perilous as it was. As I slipped and slid downward, hope and fear chewed at my guts. Maybe Ol' Kitty Guy was waiting at the truck. At seven o'clock I snowshoed into the unplowed, silent parking lot. I was still alone.

Early in our partnership, when he discovered that, after dark, I couldn't shoot the way he could run, he would abandon the hottest chase. Sometime during the homeward trek, out of the darkness, a warm muzzle would suddenly push into my hand. Because he got his sleep and I got mine, we greeted each morning with gusto. As I waited in the frosty truck, I began to think. Today had been unusual weatherwise. Up on the summit's bony ridges, laid waste by fires a century ago, hardly a day passes when the wind doesn't blow with near-hurricane force. Right now, Jiggs's tracks were still visible, but even a little eddy of wind would fill in the imprints. The sun had gone to bed in a heavy bank of clouds and I could almost taste snow. At eleven o'clock I drove home for hot coffee, then returned to wait until after two in the morning. Falling into bed for a few fitful hours, I prayed the wind would extend the truce.

When I once again reached the spot where Jiggs had left me, the grotesque shapes of the stunted spruces were barely discernible. The forest of ice was ominously still. The wind still waited. I cut a circle around the tracks, casting off sharply a half-mile to the west, thence north of Pumpelly Ridge, to finally close the ring a full half-mile from the south. I did not cross Jiggs's tracks. Somewhere within this one-mile circle above tree line lay the answer.

Yesterday's dog tracks wound up through scrubby spruces, then zigzagged along a ridge so narrow and steep that I could follow only by creeping along, peering down to make out the imprints almost two hundred feet below. On one of these round-about casts, the false ice buttress of a rock chimney loomed ahead. I inched my way around it to peer over the edge. There, 250 feet down, Jiggs's tracks went parallel to the cliff, to disappear behind a huge cornice of snow jutting out over a large boulder. The other side was clearly visible. No tracks emerged from behind the snowed-in rock. On all fours, I backed down the sheer cliff, to finally reach the tracks leading to the boulder. A heart-pounding, tightrope balancer's walk ended at the cornice. Ahead, the cliff to be traversed slanted so steeply that the mere thought of crossing it upright made my legs rubbery. Off came my snowshoes. Into the harnesses went my mittened hands. Fingers clutched the crossbars tightly. Down on all fours for better traction, I finally gained the snow chimney and looked around to its farther side. No tracks. One snowshoe crept forward, then another. Suddenly, ten feet from me yawned an opening in the snow. It was barely a foot across. The tracks disappeared into it. Not a sound could be heard. Then, as I crept closer, hoarse barking funneled upward.

Wary, lest the same fate befall me, I pressed my hands hard onto the rawhide webs and, using them for support, straddled the hole and peered down. Brown eyes, alive with love, looked up at me. Jiggs stood on the floor of a bottle-shaped hole nine or ten feet deep. Scratch marks an inch deep, as high as he could jump, furrowed the curved, ice-encrusted stone walls. He had trampled the snow that had fallen in with him into a solid mass. It was red with blood. For almost thirty hours, he had been clawing the unassailable walls, while the small opening, hardly larger than the dog himself, had provided perfect soundproofing. From only a few feet away, one could never have heard his cries.

How could I get him out without help? It would be impossible to get into the cavern without a ladder, but that meant a perilous, two-mile trip down a glass mountain, followed by an eight-mile drive home. An east wind suddenly swirled snow. Time was against me. Even if I had a rope, there was neither a tree nor a rock to which it could be fastened. What about the leather leash in my game pocket? It wouldn't be long enough, but it was strong. Swiftly, the snap hook was run through the hand loop and drawn tight enough to fashion a leather snare. Next, the lengths of rawhide thongs, used to drag out cats, were knotted together and fastened to the buckle end of the leash. Lowered into the cavern, the improvised lifeline dangled a few inches above the floor. It was long enough, but how to get the loop over Jiggs's head? He hadn't taken his lively eyes off me for a second.

My plan was risky. Once his head had been snared, the upward haul could break his neck. Could I be quick and sure enough? It took several tries before the snare settled over the upturned head. Startled, he turned, and the thong was drawn tight enough to make it secure. Realizing his life was in my hands, I pulled as I had never pulled before. Up came seventy pounds of hound. His eyes were bulging painfully when he reached the small opening. I grabbed for his collar and pulled him out. He lay on the snow. Had he been strangled as the final incident in our thirty hours of misery and uncertainty? He gasped for air. Moments later, he shook himself, wagged his tail, and bumped his head fondly against my chest. As I raised a grateful head, snowflakes stung my cheeks. It was going to be a good night to sit in front of a fire with my dog at my feet.

When April came, and my snowshoes were hung up, Jiggs turned to the family. During his first Kulish summer, as companion to a couple of horse-crazy little girls, he learned to submit to improvised martingales and bridles. His white chest and four white feet helped maintain the illusion. Because he responded so completely to their love, the girls struggled to make him human.

One day I arrived home unexpectedly in the early afternoon. As I walked by the dining room window, I glanced in. Jiggs was sitting upright in a chair at the table between two girls. A napkin had been tied around his neck. Four hands were trying to force a stiffly averted muzzle into a dish brimming with milk. He knew he didn't belong at the table; it wasn't his place. What a look of relief he gave when I suddenly stood at the open door. He didn't want to be human. He liked being a dog.

At Christmastime, each girl gave him a present. He sat beside the tree with us while the packages were opened. When his name was called out, he stood up and sniffed hard at the gay offerings. He smelled what was in them but, because they were wrapped like ours, he refused to touch them. He'd wag his tail, look at the packages, and then at us. We had to open them to show him that the frankfurters inside were really his.

He swam like a retriever, spending hours fetching sticks or chasing uncatchable beaver. He had a special place to sit in our fifteen-foot canoe, from where he scrutinized the shoreline. He realized that that's where the action was. Alert and calm, he never moved about, whether we were floating across a mirror or scudding down a five-mile-long lake with a southwest gale threatening to swamp us. He enjoyed the isolated ponds and streams we frequented. He was always the first one in and the first one out of a canoe. Whenever we came ashore, he always had an initial chore. Picking out a lonely strip of cove, he would go down to the water's edge, search it for stones, then with muzzle, mouth, and paws methodically roll and carry them into a cairn. He would fish underwater with his front paws, get them behind one and move it up to shore, where he would pick it up in his mouth, carry it over to the chosen spot, and drop it. Sometimes, when his paws weren't enough, he would duck his head, hound ears and all, to grab a stone from as much as two feet of water. Nothing distracted him until he had finished building his cairn. They were usually about ten inches high, and as much as a foot across at the bottom. Then, proudly lifting his leg, he would sign his name and depart, ready to belong to the family for the rest of the day.

When he was wet, the fur on his lower back curled. We wondered about that until Belle, his only mate, who was a registered redbone, bore him thirteen puppies. Among the ten survivors were recognizable progeny from a golden retriever, two Labradors, one Weimaraner, one bluetick, two Walkers, two redbones, and a lone Black and Tan. Everything but an Airedale. He was embarrassed by the whole terrible business and would have nothing to do with any of them. When accosted, he would head for the safety of the woods, his tail between his legs.

To me, my hounds were always guys, either "good guys" or "bad guys." Their place was in a kennel until they had earned the right to join me inside of my home. Ol' Kitty Guy was not yet four when he graduated. Allowed to come into the house to lick his wounds, he behaved exactly as I would if suddenly transported from the New Hampshire woods into Buckingham Palace. Uneasy and unsure of himself, he retired under the kitchen table to adjust to the royal surroundings. If I raised my voice, he disappeared. He never took advantage of any furniture, even when left unguarded. Venison steaks, left on a low table to defrost while we were gone, remained untouched. Even when placed on the floor in front of him, any meat still in butcher's wrapping paper would not be eaten. He figured we'd made a mistake. At mealtime, he watched us from a respectful distance.

His turn would come later, for he was the link between table and dishpan.

Gentle as a loving woman, he could also be as vicious as one scorned. When left alone in the house, he greeted strangers so savagely that, over the years, we forgot what pedlars looked like. Once, one of my customers, hastily dumped off with his duffle by a spouse eager to get back to her urban cronies, spent a fearful afternoon in ten-degree temperatures. Unable to accept our written invitation to "go right in and make yourself at home," because Jiggs didn't know him from a Fuller Brush salesman, the man who had engaged me as a guide had wandered into the garage where my saws and axes were kept. At dusk, when I returned, a mountain of wood welcomed me.

Jiggs disliked only one man. His *bête noir* lived alone, year round, in a tent, in the deep woods about two miles from our house. He had the biggest kitchen in New Hampshire, for he cooked his meals on a black iron stove, out-of-doors, with the sky for a ceiling, and the wind for a fan. How amazed I was the first time I saw a steaming pumpkin pie emerge from that oven to spice below zero air.

His only companions were a half-dozen huskies. During the winter, once or twice a week, the crack of a whip and the shout of "Mush, damn ya!" could be heard long before he appeared. As if out of the pages of Jack London, he rode the runners of a dog sled, his chest-long white whiskers stiff with frost. The sledge would shoot by with runners striking sparks as they hit the plowed road.

Jiggs couldn't stand the sight of him, but he never actually attacked his enemy. Snarling and snapping, he would run abreast of him and his team from the moment they burst into sight until they disappeared down the hill. He always kept far enough away to avoid the whip, but close enough to incite the dogs. His hatred for the sled-dog man troubled me, especially since Kitty Guy had started the feud. "Baldy" had never either insulted nor struck him. But hate "Baldy" he did. Obviously, Jiggs didn't think dogs should be forced to work. Pulling a loaded sledge was for horses. Dogs were supposed to hunt bobcats all day, in snow up to a man's neck.

While Jiggs was still young, he learned to kill snakes, and after he retired from active service, he spent hours hunting for them in fields and in stone walls, following their pungent, mouth-foaming scent. The neighbor's cats also got regular workouts, creating a painful paradox he was never able to fathom. He expected me to shoot them just as I had their wild cousins. The fact that I did not remained the one facet of my character he was never quite able to understand. As he grew older, his game became smaller: woodchucks, field mice, spiders, and finally, ants.

He never took advantage of our love. He was his most demonstrative while sitting beside me in a car. Suddenly a black head would push up under my driving arm and press firmly against my chest. For miles, he would sit thus quietly.

Wherever I went, he went. One bright fall day we waited our turn on a lift at the base of Sunapee Mountain. My wife and daughters were already exclaiming over the brilliant foliage viewed from the single chairs as they rode up the mountain. When my chair approached, I sat down quickly. "Jump!" I ordered, and up Jiggs jumped to squeeze in beside me. One arm around him, I could feel his body, relaxed and comfortable against mine. Alert and interested, he looked all around, unruffled even when our chair, rocked by gusts of wind, swung like a pendulum over granite ledges forty feet below.

In the early fall of 1964, when the Old Guy was thirteen, he ran the wilderness Allagash River in a canoe with us. Because of his age, we had at first planned to have him stay with a friend, but when it was revealed that they had a cat, our plans were changed.

"John," my wife pleaded, "he's been retired for four years. Surely he wouldn't mind being with a cat now! The war is over."

"Never!" I objected. "It would be like putting General Patton in the Peace Corps."

So down one hundred miles of sometimes turbulent river he went, wedged happily between my legs as I paddled stern. Calmly he watched the rapids in Chase Carry boil around him. On nights when we knew a half-inch of frost would coat the tiny tent by morning, we made a place for him to sleep inside, near the front opening. The unaccustomed intimacy made him uncomfortable, for he never forgot that he was a dog.

In his fifteenth year, we realized that Ol' Kitty Guy was stone deaf. A tenuous film began to form over his bright eyes. His hind legs were hard pressed to support his body. We had to carry him up and down stairs, but he still showed his respect, even at a personal price. Whenever he was left alone in the house for hours, as he sometimes had to be, he greeted us with happy thumps of his worn tail, but the greeting was necessarily short. His love seemed to grow stronger as his body grew weaker. He didn't seem to mind being old. As naturally as we were eager to give it, he accepted the help he needed to get up. God designs us to grow old, man and dog alike. It is love that gives our mutual fate its final dignity.

Perhaps an animal does not know it is mortal until close to the end. I think he had come to understand and to accept. It was I who would not yield. We hadn't truly realized how he had aged until our younger daughter returned home after two years in Colorado. When she walked into the living room and saw him sleeping in front of the fireplace, his wasted legs and feet moving as, in his dreams, he once more coursed over Osgood Mountain, her face crumpled. "Oh, Dad," she wept, "he's *old.*"

Because he was unable to take care of himself, many friends, kind and good people though they were, had asked why I didn't have Jiggs put to sleep. How could I execute the truest friend I had ever had, one who had loved me since the first moment we met, and who never once had given me cause to be angry? When I had needed Jiggs, he had never failed me. Now he needed me. I wouldn't fail him. So I had sworn that, unless in pain, Kitty Guy would die naturally, either in his sleep or in my arms.

Now I sat on the floor in front of the fireplace, but the blazing hardwood logs couldn't warm me on that cold morning in May. Outside, it was snowing. I held the Ol' Guy's head and shoulders in my lap. He seemed more comfortable that way. All night we sat in this embrace. Now I waited for the doctor to come to ease the Ol' Guy's now deadly pain, and to increase my own.

It took me a long time to make his coffin. It had to be big enough to hold his bedding. I wanted his tired body to rest stretched out on his favorite rug. A weeping daughter sat on the cellar stairs and watched the pine box take shape. Her gentle, understanding husband guided my fingers that held the nails I couldn't always see. Together, all four of us dug the grave under a towering oak on the hill behind the cabin. Somehow, the carpentry and the digging eased the pain.

He left me as he had come to me—in a snowstorm. It was fitting that I buried him while the soft, white flakes drifted into his grave. Each year, he and I had come alive with the snow. The part of me that only he had known went with him on his longest hunt.

12. Beaver: Benefactors of the Boondocks

Beaver are the salt of the animal earth. Designed by nature to be peaceable and persevering, they are also endowed with creative genius. Ask any distraught small-town road agent, fighting to keep dirt roads from being continually inundated. Beaver do not know how to quit being beaver.

Zoologists have allocated to Sir Flat Tail the seventh rung on the ladder of animal intelligence. From what I have observed in the woods, the beaver has gnawed its way up at least a couple of rungs while the scientists weren't looking.

If intelligence can be defined as "the shaping of the present and the future by the past," consider the dilemma of part-time trappers, with full-time reputations, who have complained to me about not being able to trap out an entire colony.

"What's wrong, John? The first four or five are easy to catch. Then no more. It's always the big ones that are left. All they do is eat up the bait."

One ten-generation Yankee gentleman trapper was using a standard water set, whereby a dead pole long enough to extend above the ice is thrust into bottom mud. About a foot above the base of the pole, an exposed trap is set on a "platform" of two sticks nailed into the wood. A couple of feet above the trap, poplar sticks are wired to the pole.

"John, the next day the bait is gone, so I nail a second platform on the other side, and wire lots of bait completely around the dead stick. The next day, the bait is gone again!"

Grandpa beaver had observed what had happened to his family when they had stepped onto the "platform." Still, that poplar looked mighty good. He hadn't had any for weeks. He swam over to the pole, turned upside down into diving position, sidled down, approaching the fresh bait from above. Treading water, up-ended, he circled the stick, relishing every last nibble.

"Tell me, Andrew," I chuckled, "if you were going to rob a bank, would you enter through a door where uniformed policemen were standing, brass buttons gleaming?"

The intelligence of these amphibious engineers is matched only by their unflagging diligence. No current is too swift, no tree too tall, no ridge too steep, no canal too long, no dam too demanding. It has been estimated that sixty million beaver gnawed at the primitive forests of North America before the Mayflower landed. That ship's cargo included the age-old struggle between money and morality. During the very first meeting with Samoset, the Puritans, after oblique glances at the lustrous pelts worn by the red welcoming committee, talked business. Thanks to the beaver trade, by 1640 the Massachusetts Bay Colony was debt-free. When the white immigrants made the beaver pelt the unit of exchange, the red natives traded them for food, clothing, and firearms. The Indian was not intent on building an empire. He already had one.

By 1690, licensing systems regarding the sale of beaver were in effect in every colony except New Hampshire. For the first and last time, that staid state was wide open. The territory had been given to loyal supporters of the King of England in the form of land grants. The size of a tract equaled the depth of a loyalty. These boys made no bones about long-run consequences. Money was what they were after, and they found it in ponds and streams, on the backs of thousands upon thousands of beaver.

Towns were settled and roads built close to beaver ponds. Cattle grew fat on wild grass growing in meadows kept fertile and moist by these backed-up ponds, some of which covered several hundred acres. For years, even after they were abandoned, the dams still served as foot bridges for crossing brooks and ponds. By 1820, the state was nearly trapped-out. By the end of the century, even in the most primitive parts of New Hampshire, beaver were hard to find. In 1905, the legislature finally gave the dam builders complete protection. It was too late. There wasn't a beaver left. Again, a barn door was locked after a horse was stolen. Drawing blood with their spurs, the thieves had sped westward.

Sometime during the next decade, beaver swam across the borders into New Hampshire from Maine or from Canada. By 1926, the northernmost counties once more buzzed with the sound of wood-chipping teeth. By 1940, fifty-eight of these wood choppers were reintroduced in the southern counties, where they have since flourished, oftentimes to the point of frustration for farmers, road agents, game wardens, and the competing (two-legged) lumberers. Only the beaver remain unshaken by their population explosion.

In nature's ledger, the beaver is listed in the credit column, for she designed it to benefit and to enrich, not only other animals and birds, but all wildlife. Lining man's pocketbook was an unforeseen by-product.

A new beaver pond is the first and the strongest link in an ever expanding, far-reaching chain of events. Even when beaver, intent on building bigger and better dams, abandon ponds, or when an entire colony is trapped out, the legacy they leave is as long-lasting as the dams they have built. A beaver pond lifts the horizons of the entire forest community. Within months, it becomes the civic center for inhabitants from miles around; a forest plaza, with varied restaurants; the most popular meeting place for animals that enjoy good food. Fish thrive and multiply. Turtles settle in rich mud. Water snakes grow thick on frogs and salamanders. Mink dart about the shores, competing with snakes for fish and lizards. Otter, who previously may have visited the barren brook but once a year, now include it on their regular itinerary. Muskrats come to eat juicy water plants and stay to build homes. They sulk when raccoon, intent on frogging and salamandering, end up trying to muscle in on the muskrat's supply of mussels. The barred and the great horned owls watch the contest from seats in the mezzanine. Sparrow and marsh hawks hover overhead, contemplating the menu.

When the inundated coniferous trees die, many of them a foot or more in diameter, insects lay eggs under loosened bark. Soon, woodpeckers, flycatchers, and cedar waxwings flock to the waterfront snack bars. The deciduous trees die and dry on the stump, furnishing the finest of firewood for any man willing to be twice warmed. Pioneer trees—gray birch, pin cherry, poplar, and swamp maple—as well as a variety of low-growing shrubs, spring up in the "chopped out" areas around the pond. New shoots begin to grow from hardwood stumps cut by beaver. The tender stems from birches, maples, and even from oaks, supply gourmet food, not only for beaver, but for deer, hare, and partridge. Wildlife runs, hops, and flies to feast on ground shrubs and their fruits. Bobcats begin to hunt the area regularly. Their yellow eyes, in "quick draw" holsters, hope to zero in on a varying hare or on an unarmed beaver kit lumbering alone on a hillside. Foxes lurk around the marsh, their noses peeled for a careless muskrat. The decaying leaves from felled trees enrich moist humus. Earthworms pop up everywhere, luring woodcocks from miles around. Waterfowl—black ducks, teal, mallards—and wood ducks bob for burr, smart, and duck weeds. They stay to raise broods in comparative safety. Stilt-legged bitterns and herons stalk the shores, to suddenly stand still while great, glittering eyes probe for frogs, snakes, lizards, and tadpoles. Shorebirds—the greater and lesser yellowlegs, sandpipers, and the northern water thrush—jerk along the shore seeking shallow-water insects and minute plant life. From afar, plants, foreign to the area, are introduced by migrant waterfowl. (A beaver pond in Stoddard, abandoned for a decade, harbors the only specimen of *Phragmites communis,* a stately tidewater plant, I have seen inland. Its graceful plumes, a dozen feet high, have learned to bend to winter winds.) As a natural bonus, scrappy, native trout await for those human fishermen who still remember how to walk.

In a beaver pond, life pulses everywhere. In spring, summer, and fall, it splashes, hums, chirps, peeps, bleats, barks, croaks, quacks, hoots, and sings. In winter, one cannot hear the pulse beats. But if one knows where to look, he can put his finger on any one of a dozen places and feel the throb of its life.

The damage beaver cause by flooding dirt roads, timber lots, and cultivated meadowlands is negligible compared to the immense contribution they make to the wildlife community. The problem lies in the fact that the injury touches man's most sensitive nerve: the one connected to his purse strings. Some biologists claim that beaver ponds might disrupt migrations of spawning fish, but I have yet to see a beaver dam a female trout couldn't get through.

In 1948, after the state of New Hampshire had had to pay damages for too many backed-up beaver ponds, the "powers that be" began to try to figure out how they could have their cake and eat it too. As a result, for almost twenty years, the standard technique for managing the dam builders has been through the use of ordinary, four-inch, perforated, fiber drainpipe. The initial experiment resulted in a duel of engineering talents. Two sections of pipe were extended through a dam. That stopped the flooding until the resident engineers found the open, upstream end and proceeded to plug it. Human technicians countered by suspending the pipe from hardwood posts, to which it had been tied with baling wire, after hand-plugging the upstream end with a wooden disc. The flat tails reciprocated by chewing down the posts, thus forcing the two sections apart, then filling up the break with sticks and mud. Man lunged back to replace wood with iron stakes, nailing pipe sections to reinforced couplings, always keeping the upstream end plugged. At long last, the human duelist won. The perforations in the drainpipe maintained the desired water level, and the beaver could neither remove nor sabotage the final arrangement. In some dams, barbed wire around a central pipe helps to maintain a "no-beaver's-land." The state of Maine uses corrugated pipe in a device called a snorkel, which allows water to run through a dam, thus preventing flooding of negotiable assets. It always hurts me to see dynamited dams.

The beaver, averaging thirty-five to forty pounds in weight, are the largest of North American rodents, second only to the South American capybara. (Judging by appearances, the ancestors of both must have shared a saurian swamp.) It was not unusual for me to catch senior citizens weighing up to seventy-five pounds. The largest beaver recorded was a monster trapped in Wisconsin in 1921 that weighed 110 pounds. When the Indian was in charge of America's game management, hundred-pound-plus beaver were common. No matter how large, these amazing creatures cannot be described as full-grown, for beaver never stop growing.

Like a man who swings a shovel or an axe for a living, the animal's body is solid, tough, and muscular. Its back is humped, its head big, stately, and Roman-nosed. The distance between a swimming beaver's ears tells me its size and generation. Age and experience broaden a grandpa's skull. The jaws are the animal's powerhouse. Each of them holds two, curved, cutting teeth in front. Portable, self-sharpening sawmills, they can cut down trees as much as four feet in diameter. In the late forties, in Stratton Pond, Vermont, I marveled at the largest beaver-felled tree in my experience. A tremendously large poplar, at the cut-off point, it measured twenty-eight inches in diameter and was over sixty feet in length. Top beaver lumberjacks hail from British Columbia, where they are reported to have dropped an aspen over four feet through and 110 feet tall. The four incisors never stop growing; they just wear away. Segments of a circle, if, by accident, they break off and cannot be used, they curve back, growing into the skull and killing the animal. There are no lazy beaver.

Besides teeth that never wear out, these civil engineers have two pairs of eyelids, one used only for underwater swimming. Like a scuba diver's goggles, they are transparent and protect the swimmer's eyes from sticks and floating objects while affording a clear line of vision.

When designing the beaver, nature was in a generous mood. When it submerges, not only do goggles go on, but ear and nose valves close automatically. (The hatches on nuclear-powered submarines still have to be closed by intent.) Moreover, special mouth flaps meet behind the incisors, like the flaps on a tent, making it possible for the animal to chew wood underwater without ingesting either unwanted slivers or pond water. It moves me that the beaver himself does not realize what special gifts he has been given. When he is lumbering on a hillside and spots a hare hopping toward him, he doesn't drop his axe to shout, "Hey, Peter, come over here. I got something you haven't got!"

The animal's front feet are small and shaped like those of a squirrel. Like human hands, they are used for eating, for working, and for carrying. When swimming, empty paws are held up against stalwart chests to lessen water resistance. This doesn't happen too often, for every beaver is taught by example never to waste effort by entering a lodge or by approaching a dam empty-handed. They dive to the bottom, scoop up mud, roll it into a little football, clutch it like a professional quarterback, and, swimming with their great hind feet, head for the construction area. In order to eat, they use the front feet to hold a piece of wood in the way we do when gnawing an ear of corn, or as a musician does when playing a harmonica, or as I do when devouring a squaretail trout. Sticks and logs are carried with the front feet or with the jaws. Although I have never seen them do it, judging by the size of logs and stones in dams, I am sure they must have to "bulldoze" some of them, using their shoulders and backs as pushers.

Big and webbed, the beaver's hind feet are stern paddles, propelling it with steady, powerful strokes. Attached to each hind foot is a minor miracle. Let me tell the whole story. In March, 1946, during an eleven-day season, I trapped my first beavers. In 1944, while I was in the South Pacific, the first season had been declared in New Hampshire's southern counties. I minutely examined each animal I caught, with the knowing eye of one who already understood the wonders of the river otter. I studied the sturdily beautiful, oily-haired, always immaculate creature with growing wonderment. The hind feet particularly intrigued me. I pondered a strange phenomenon. I had never seen anything like it before. No encyclopedias mentioned it. Other trappers shrugged their shoulders. Most of them are interested in "how much?" rather than in "why?" By the end of that first season I had come to a conclusion. Nature rarely adds useless extras to her models. The split toenail on each of the second toes was no mere accessory. It was used every day. I called it a pocket comb, a preening tool, which the beaver uses to clean out cockleburs and to groom itself. The engineers in beaver ponds do not have white-collar jobs. Each morning, when tired and dirty, as they return home from work, one might say to an equally work-worn buddy, "How about holding the mirror for me, Joe? Then afterwards, I'll hold it for you." After the cleaning and the preening, comes the grooming. Ejecting a dab of viscid, yellow castoreum hair cream onto a hind foot, the beaver rubs it in all over its body. The special comb can reach every hair of its fur coat. This is why all beaver have a shiny, waterproof coat that smells pleasantly musky. I have never found a flea, a wood tick, or a cocklebur on a winter beaver. I wonder whether in summer, they have as hard a time as I, trying to keep ahead of insects.

The tail of a large beaver averages about a foot in length, and six to eight inches in width. Hairless and lizard-textured, it is oval-shaped, two to three inches thick where it joins the body. It tapers down to almost a double layer of skin at its tip. I am sure the first canoe paddle was fashioned with one glittering black eye on a beaver's tail, for the American Indian had seen what it could accomplish in water. The tail performs a threefold function. The major one is that of both rudder and diving plane. Swimming is a beaver's chief occupation. When diving or emerging at a sharp angle as well as when dragging heavy logs or bulky branches, its tail acts as a counterforce. Remember, much of the time the animals must swim against the current. I am convinced it also uses its tail as an auxiliary source of power whenever an extra spurt of speed is needed. Many times, I have watched a king-sized beaver shift into overdrive. The tail begins to undulate through water, producing a forward thrust which is based on the same principle which operates when a fisherman sculls a dory.

The beaver's tail serves as a defensive warning signal as effective as any man has devised. Many a fisherman, daydreaming while taking advantage of an evening trout rise, has been startled back to reality by the resounding slap of a beaver's broad tail on water. It is one of the loudest sounds made by a wild animal. I have been jerked from a sound sleep by the crack of Ol' Dad's tail as it slapped the surface of water a hundred yards from my tent. "Take cover! I smell a cat," or, "Send the children back into the lodge. A great horned owl just flew over the pond."

The tail's third function is that of a steadying "platform" when the animal is felling trees. Like a front-end loading or a ditch-digging machine, a beaver needs a solid surface from which to perform its oftentimes dangerous duties. Many of the beaver I trapped had healed-over holes or notches in their tails. Some of the old-timers had several of each. One whopper had a forked tail; three-quarters of it, a pie-shaped wedge, was missing. How did that happen?

Lumbering is hazardous work. When cutting down a tree a foot or more in diameter, beaver first girdle the trunk, chewing around it in a spoollike fashion, until the trunk resembles an hourglass. Finally, a single bite is all that is holding erect a thousand pounds, or more. If the tree slants to either side, it falls cleanly. If the tree stands straight, but wears a lopsided crown of branches, it falls cleanly. But when the tree is perfect, standing straight and tall, with symmetrical branches perfectly balanced, catastrophe can result. Unless a nocturnal wind is blowing hard enough to produce a side force, such a tree will suddenly let go, slip down along the lower spool of the hourglass formed by the immovable stump, and ricochet. If the tree crashes to the ground, the chopper may still escape with only a bruised ego. But if the tree is hung up by others around it, the thrusting point of the spool nails the awkward, unsuspecting lumberer to the ground, often through its broad tail. Blood spatters over wood chips and leaves. The earth is churned up, as the victim, shifting into four-wheel drive, digs deep into humus. Slowly, the tail begins to rip. After a struggle that may last into daylight hours, a large beaver can usually tear itself free. During long winter evenings in a snowbound lodge, he can tell his family about his "moment of truth." He has the tail to back him up.

Other trappers have told me they have caught beaver with damaged tails. "Boy, John, what a fight it must have had with another beaver!" Beaver do not fight among themselves. Males do not have harems; they are monogamous. Sometimes a pinned beaver cannot break away when speared by a tree. Although I have never found one who had given his life for his colony, I am sure many make the supreme sacrifice. A dead beaver does not lie around long. The meat is relished by cats, bears, foxes, and the great horned owl. Squirrels and mice devour the bones. There are no leftovers in nature's commissary.

13. Mother Nature's Corps of Engineers

Every creature on earth has been given a special characteristic with which to protect itself so that it may continue on earth. The same unique trait, directly or indirectly, helps the animal to find its food. The ability to harness and to control water is the gift nature has given the beaver. If it could not escape into deep water, it would not survive for long. No beaver can run; it waddles, dragging its tail like an anchor. As it is, even if its worst enemies enjoyed swimming, they would never catch the Flat Tails. Bears and bobcats don't know how to submerge. They were designed without conning towers.

Civil engineers do not build dams just anywhere. Neither do beaver. Miles of brooks are surveyed, acres of woodlots cruised, before a site is chosen. Any dam builder worth his poplar bark degree, picks a site as judiciously as any Army Corps engineer. (Beaver have a distinct advantage: they don't have to keep an anxious eye on the Secretary of the Interior.) The determining factor in making a choice is the food supply. The bark from poplar trees is the most desirable. In descending order are the bark from pin cherries, white, gray, and yellow birches, and alder. In other parts of the country it seems that beaver rarely eat alder. Swamp Yankee beaver do.

In the spring, when a pair of mature beaver emigrate up a brook, the water may be no more than four inches deep at the chosen site. Under these circumstances, how can a fifty-pound animal hide? Where can it go? The ensuing nights of untiring toil are the most dangerous the dam builders experience. Once the dam is high enough to form a pond, they are safe. Enemies approach, but animal enemies don't hang around hour after hour, night after night, like hired killers. Hunger motivates natural foes. When it finds that its supper is safe in a pond, the cat or bear shrugs its shoulders philosophically, and goes away to look for something else. After a while, the beaver come out to begin logging. They start by cutting close to the water. Not only because the trees are there, but because, until the loggers get the lay of the land, they want to be able to silence their saws and to jump into the water at a moment's notice. All the while they are learning about their surroundings, they are also getting to know the habits of their enemies. If the next links on nature's food chain aren't too persistent, the lumberjack engineers venture farther and farther into the woods away from the pond. I have rarely found them cutting more than 300 feet from water except when they build a lodge in the bank of an island on a pond or a lake, or whenever an island is formed in the pond they have established. Then they seem to feel as secure as a feudal lord in his castle, surrounded by a moat. Here, the beaver picks out trees at random anywhere, and sometimes even cuts off the entire island. The law of averages works for man and animal alike. An occasional overconfident kit is scooped up by a great horned owl or a fox. Sometimes, even a graduate engineer makes a mistake and comes face to face with the business end of a bobcat.

A beaver dam is an engineering marvel. Small wonder our greatest technological university has a busy beaver for its insignia! These construction geniuses understand water currents, water pressures, and their relationship to the bordering land contours. No matter how big the supply of luscious poplars, if the brook flows through a valley too steep-sided to afford the space for a reasonable-size pond, the civil engineers don't even bother to set up their transits.

We all know that beaver use logs, sticks, mud, and stones to build dams. This is still true of the backwoods provincial, but times are changing. What about the beaver exposed more and more to civilization? Their technology also is becoming increasingly sophisticated.

Most portable sawmills are set up near water, usually a brook. The mill and its operators must have water. The lumberjacks live in nearby tarpaper shacks. When the logging job is finished, the men move out, leaving the flimsy shanties. Varied junk is left behind, scattered over the landscape. While exploring a beaver dam built a few hundred yards below an abandoned log job, I saw that the resident engineer in charge of dam construction had taken advantage of a nearby windfall. In the dam was embedded a rusty bedspring, its steel coils interwoven with sticks and mud. How I wish I could have been there the night they had carried it seventy-five feet down to the water.

Another dam sports an old bicycle frame. It is embedded upright among sticks, stones, and mud. (I wonder what they did with the wheels.) During the last decade, more and more, I find beer cans, discarded by fishermen, stuffed into crevices and packed with mud. Any M.I.T. graduate understands the merits of steel reinforcements. Someday, I'm going to try leaving a bag of cement and a trowel at a construction site.

Most beaver I have observed work on a dam and a lodge concurrently. Lodges are usually built out in the open, completely surrounded by water. The size of a lodge tells me the size of a colony. A tenement block usually houses more people than a trailer. As the size of a family increases, ells are added and ceilings raised, as more loft bedrooms are needed. I have seen several lodges ten to eleven feet high, housing a colony of at least twelve. It usually takes four or five years to produce such a clan. The largest pond I have seen in our area was one on Moose Brook in Hancock, in the fifties. Almost a mile long, with some coves a half-mile wide, it contained four tremendous, super-duplex lodges. It took the inhabitants twelve years to eat up the existing food supply.

Once in a while, a building genius outdoes himself. In the late forties, while scouting for fur, I came upon a lumbering operation, terminated four or five years before. The human lumberjacks had built a shed near a brook. Haphazardly constructed out of old lumber, but with a tarpaper roof, it had been used to house horses and to store harnesses and tools. A few years after the logging was finished, beaver had come to the brook. They built a dam several hundred yards downstream. Soon the entire area was flooded so that water surrounded the old shed.

My first glimpse of the pond was one of amazed disbelief, followed by a shout of delight. Guess who had built a lodge within a lodge, complete with the usual underwater entrance? The cracks between the old boards bulged with sticks and mud. Later that fall, I was guiding a deer hunter, who was also a close friend. Because he was in the federal Fish and Wildlife Service, I went out of my way to show him the "rural redevelopment."

Nonplussed, he stared at the beaver-made lodge inside a man-made shed.

"Why in the world would a beaver want to build its house in there?" he asked.

"That's easy," I replied. "Now they can work on their lodge when it's raining without getting wet."

Beaver do not hibernate. Suddenly, one frosty night in October, as though a button had been pressed, their activity cycle speeds up. No longer are the nights long enough; they do not wait for dark to begin working, nor do they quit at daybreak. Until the ice comes, every member of a beaver colony works overtime. Icebound during four to five months of New England winter, they must store a winter's supply of food before the pond freezes. How do they know it will be four months before they see Orion again? Because beaver do not leave their pond during winter, nature warns, "Get that pantry full, or you will die!"

When the first frost nips a man's fingers, and sends him scurrying for storm windows, the entire beaver colony begins to spend all night, every night, chopping down trees, cutting them up into manageable sections, and dragging the cordwood to water's edge. From there, they swim near the lodge with the winter's victuals. A short distance from the entrance, where the water is deep, they dive to the bottom with their poplar, cherry, and birch, the bark still on them. To keep the wood from floating, one end is stuck into the mud. As the weeks progress, the food pile reaches near to the surface. The beaver realize that any part of the brush reaching above water is wasted. It will be beyond reach when the ice comes. Exactly as a farmer extends the stacked cordwood for his winter fuel supply, the beaver elongates his food pile underwater. The communal effort continues tirelessly until the freeze-up. During this busiest season, the average beaver pond resembles a freeway during rush hours, with too many drivers and not enough rotaries and traffic lights. It surprises me that they have not yet developed a corps of traffic engineers.

During winter, at mealtime, beaver leave the lodge via an underwater tunnel, swim over to the pantry, cut themselves a stake, and swim back up into the dining room to eat it. The food supply, cut and stored by mid-November, stands in water throughout December, January, February, and on into March. The bark begins to sour. It smells and tastes unpleasant, like food we may keep in the refrigerator too long. It doesn't make one sick; it's just difficult to swallow. Weeks before the vernal equinox balances the heavens, beaver are dreaming of fresh bark.

By mid-March, in our area, snow lies heavy and packed on the ground. Nights are cold, but already the sun is high. Somewhere around the edges of most beaver ponds, a spring has bubbled away the winter under ice. Long daylight hours with a waxing sun melt its thinly frozen lid. Soon, a hole appears. Then, the spring is free. Beaver, waiting for a door big enough to squeeze through, break out of prison to clamber ashore. Pure air, fresh food, and if it's the right time of the month, a full moon. Perhaps, desire overcoming caution, one might venture out during daylight hours, but I happen to have seen only those taking moonlight walks. Waddling over snow-covered ice, they head for the woods. After cutting down a fresh sandwich that is usually small enough to handle in one piece, they drag it back to the spring hole. If it won't fit through intact, they cut it up into bite-size pieces, shove them through, then swim back to the lodge to feast. Tempted as they might be, they do not eat their sandwiches on land. Bobcats are rutting. One cat is a risk; a pair can be a catastrophe.

Beaver are sanitary as well as civil engineers. I have never found their scat inside or near a lodge. They defecate in a particular part of a pond, always using the same area. It is usually well away from the house. As you would suppose, their scat resembles pressed wood. Because it disintegrates so readily, it is not easily discernible, even in shallow water.

Besides dams and lodges, beaver design and build logging

roads and cargo canals. When the immediate perimeters of a pond have been lumbered off, and scouting reveals desirable food in the "out-back," these engineers lay out and dig waterways. The channels are deep enough for them to swim in without touching bottom. The longest one I have seen was engineered 150 feet to a pond. When trees growing in the canal are not the right flavor, beaver don't cut them down; they sever the roots, so these do not interfere with timber being conveyed down a "chute." New Hampshire is not good canal-digging country. Whenever a little knoll six or seven feet high interrupts a water channel, the trail continues overland, and thence back into the canal. I have yet to see a tunnel dug under such a rise of ground. After a few hundred more generations of evolution, we may find lock chambers in all beaver canals.

A few miles from my cabin, a colony of beaver have built a series of baffle dams, two to four feet high, on a tiny brook which is the outlet to a large pond the colony established five years ago. The beaver still live in the mother pond, but, each night, they travel downstream to work on the new construction site. The brook's flowage was so meager that swimming was impossible. Top engineering brains were picked. Following expert advice, the builders resolved the problem by digging out the center of the tricklet, piling the stones and gravel onto either side, and thereby forming a channel deep enough for even a great-grandpa to swim in without scraping his toes. I have yet to see a better job of dredging.

Wherever waterways are not feasible, beaver build logging roads. If a surveyed route is blocked by a stump or by a fallen tree, a right-of-way is cut that inevitably results in the shortest, straightest distance possible to the pond. I have seen such rights-of-way cut through hurricane-felled pine trees, two feet in diameter. The beaver chewed them through, then, like bulldozers, they pushed and shoved the huge sections out of the way. Many times, I have walked down their heavily traveled log roads three hundred feet to reach the pond.

Dedicated drudges that they usually are, even beaver get spring fever. Mature males emerge from a rough winter under ice with lackadaisical hearts. Shingling the roof and reputtying the windows is the last thing on their minds. Mrs. Beaver, preparing for her confinement, is poor company. An uneasy, inner restlessness, a mate's ill temper, and the stirring chant of peepers, quickens the beaver's resolve. Open brooks and rushing rivers look tempting. It is vacation time in the colony. Some males spend it cruising up and down a river for a week. Others will not return for a month, while a few adventurers will be gone until frosts remind them that, back home, there is work to be done. Those beaver who have reached marrying age leave, some never to return. They go to establish a dynasty elsewhere.

Truant beaver communicate with one another as human vagabonds often do, by building cairns. Wherever a projection from a ledge, a large flat stone, or a big log sticks out above water, a roving engineer comes ashore bearing mud, leaves, and sticks. From the debris he fashions a cairn. The finished landmark, usually eight to ten inches high, and as much as a foot across, is crowned with a dab of personal perfume, castoreum, the most precious of a beaver's possessions.

A few days later, another itinerant swims up the river. When he sees the cairn he goes ashore to look it over. "Ye Gods," he exclaims, "Uncle Joe was here a few days ago. I haven't seen him in years. He must be upstream."

Meanwhile, back at the lodge, Mrs. Beaver has delivered her young. She nurses her kits for about a month. Within a few days after birth, the babies take their first swim. They need no urging, but go on their shakedown cruise like old salts. Within a few weeks, the yearlings, banished during the delivery, are allowed to return to the lodge. Soon, father, refreshed and as light-hearted as a beaver knows how to be, returns, full of stories about adventures while abroad—stories to be shared with wide-eyed progeny in the leisurely summer months.

To me, the most remarkable characteristic of this remarkable animal is the fact that the kits remain with the parents for two full years, before filial ties are severed. It is not difficult to understand nature's logic. What would happen if Mrs. Beaver, like Mrs. Bobcat, Mrs. Fox, or Mrs. Mink, ousted her young after a few months? How would beaver learn to build houses snug enough to withstand New England winters, or how to store food, or how to construct dams, or how to dig canals, or how to fell trees? A complex society demands a complicated education, or is it the other way around?

The young stay with father and mother for the same reasons that aspiring engineers matriculate at technical schools. The civil engineering curriculum of the apprentice beaver includes courses in Introductory Tree Identification, Advanced Selective Cutting, The Natural Philosophy of Water Pressures and Currents, Scientific Erosion Control, Tunnel- and Canal-building, Sluicing of Logs, Types of Mud and Clay, and Stream Flow Characteristics. In addition, "city" beaver must take graduate courses in Trap Identification and the Utilization of Metals in Dam Structures.

The beaver's ingenuity was demonstrated to me one season when I was running a trapline about fourteen miles long. Each day, on snowshoes, I covered a route consisting of a series of isolated beaver ponds. Trapping these animals involves using one's snowshoe as a shovel to clear away snow, then chopping a hole in the ice with a steel chisel. After rolling shirt sleeves and long underwear above the elbow, one reaches into the cold water to set a trap. Because it was not waterproof, I wore my wristwatch fastened to my shirt collar by running the leather strap through the topmost buttonhole and buckling it. One day, well after dark, returning to my little truck, I reached up for my watch. It was gone! Perhaps I had forgotten to attach it to my collar that morning? Trouser and shirt pockets were pulled inside out. I must have lost it, no doubt on some beaver pond while lying on my stomach, looking down into the hole I had chopped. It was a valuable watch. For twenty years, we had shared every moment. Besides, it had taken two years of scrimping during the depression to save up enough money to buy it. I wanted to find it. It was too late that evening, but at daybreak I was back. Until dark, I retraced the fourteen-mile circuit. There was no time to tend traps. Every likely looking hole in the snow where it might have fallen was scrutinized. The watch was gone for good. Dejected, I gave up my search and tried to forget about it.

The following June, I made it a point to fish the same string of beaver ponds for trout. I noticed that beaver were lumbering across the pond. My new watch said it was exactly 6:00 A.M. when the logging operation ceased. Across the pond came seven beaver. Swimming in single file, they followed a leader whose head was twice the size of any of the others on the work crew. Upon reaching the lodge, they submerged, one after another.

Trout do not rise for flies during the day, so I, too, took a siesta on a sunny bank. I wanted to make sure I would be there for the evening rise. An hour before sunset I began to cast again. At exactly seven o'clock a stately head emerged in front of the lodge, followed by six successive heads of various dimensions. Purposefully, they swam toward the farther shore, to disappear into the twilight. Within moments, the night shift had resumed logging.

As I dismantled my rod, I began thinking . . . 6:00 A.M. to the minute . . . 7:00 P.M. on the dot . . . The boss beaver must be wearing my watch.

14. Otter Observed: Best Hearts, Best Brains

A wet black torpedo exploded, spraying water and slush. A hundred yards from shore, nature's own pop artist hung in midair, drops of water rolling from the curved whiskers, the thick forelegs held flush against the belly. The wide mouth gripped a struggling fish. Bobbing its tapered head in a series of gleeful jerks, the otter spun the fish around until the head pointed "down the hatch." Breaking it up with loud crunches, it swallowed its catch and disappeared under the slush without a sound. Within minutes it burst up again, fifty feet away, with another fish. Each time it repeated the spinning ritual before swallowing. That otter popped up all over the cove. It did this with such zest for the very art of living, I still can't decide whether it most enjoyed the fishing, the spinning, the eating, or the popping.

After thousands of hours spent studying the river otter, observing them in their natural habitat still thrills me. This was one of my most delightful observations. It occurred after a heavy, early-winter snowstorm had covered a big cove in a primitive pond with slush ice. Approaching through a spruce thicket, I saw dark, round holes scattered about in the floating white mush. The cove looked as though a dozen giant doughnuts had been cut from it. At once I knew that an otter was there, that it was fishing, and that it was leaping up through the soft slush to eat its dinner. I sat down behind a heavy screen of conifers to watch the ballet through my binoculars. Never have I observed a more bewitching exhibition of choreography!

Knowledge and understanding seldom come easily or hand in hand, whether one is studying medicine, law, or nature. It has taken me nearly half a century to learn what I know about the river otter, and I learned most of it by trapping them.

What I now have in my head and in my heart was well worth the chilled fingers, the bloody feet, and the tightened belt. Brave and alert adversaries, otter lost only because I learned to think like them.

I learned to respect their intelligence, then to admire them for their character, and finally to love them for the rare and noble creatures that they are.

As a man, I would like to be the kind of man an otter is as an animal. To him, life is an adventure. He is rarely foolish, but neither does caution dull his high spirits. He doesn't try to be something he isn't. He respects himself, his wife, his children, his friends, and his professional ability. He even respects his relatives. He is loving and not afraid to show it. He is never afraid to learn something new. He is invariably good-natured and enjoys at least one belly laugh each day. He hates no one, even if they don't know how to swim or like to eat fish. A peace-loving animal, he fights only when his freedom is threatened, and then he is willing to die, fighting for what he believes in.

Many years ago I was trapping muskrat on a branch of the Millers River. At that time, I knew nothing about otter. Neither did almost anyone else, except that their dark pelts sold for upward of sixty dollars apiece.

It was mid-November and a light snow had fallen during the night. I approached a broken-down dam, where a hundred years ago, a gristmill had stood. Unfamiliar tracks went up and around it. A drag mark, from eight to ten inches wide, and three to four inches deep, with an occasional, blurred footprint on either edge, confounded me. It looked like the kind of track an alligator would have made. But this was north central Massachusetts! These were the first otter tracks I had ever seen.

I didn't know where this animal had come from, or where it was going, or why, or how. I realized it wasn't satisfied with one pond as the muskrat or the beaver are. This new animal was here; then suddenly it was not.

To a boy whose struggling parents disapproved of his interest in animals, bringing home a sixty-dollar pelt might help to relieve some of the parental pressure. Perhaps he could remain both in their good graces and in the woods. During the ensuing winter I learned all I could. The encyclopedias told me its genus, size, color, and physical characteristics, but nothing about its character or behavior. What makes an otter an otter? I could not ask any other trappers, even though they were twice my thirteen years. They knew less than I did. Only the otter could teach me what I aimed to know.

Without realizing it, I then dedicated myself to a lifetime of finding out everything I could about these enigmatic creatures. I have learned much. In many ways they still baffle me, as they baffled Henry David Thoreau. In the 1850's he wrote, "Here is an animal the size of a human infant that lives on the Sudbury River, and no one knows anything about it."

I did not take otter from their environment into mine; rather, I took myself into theirs. If I had studied one in captivity, I would have learned only how it responded to a human being in his home. How did otter treat otter in theirs?

The average adult weighs approximately nineteen pounds. For years I weighed each one. The heaviest tipped the scales at thirty-two pounds and stretched six and a half feet from nose to tail. The smallest weighed less than eight pounds. The length of an average adult is about fifty inches.

Shaped like a torpedo, the otter has a streamlined, flexible body with a tapered tail which makes up one-third of the animal's overall length. Unlike any other animal's, that appendage is triangular at the base of the spine, and tapers to a blunt point. Its shape indicates its function not only as a rudder, but as an auxiliary engine when fast pickup and power steering are needed. The skin stretches taut over a muscular body, kept trim despite a prodigious appetite. Even though the animal eats constantly, its activity is even more continual. It is always on the go, definitely no kin of the "chaise lounge" set.

Hyperactive as they are, otter require little rest, seldom sleeping more than an hour at a time. Actually, it may be even less. During summer and fall, if they are in true wilderness, they will occasionally "make a nest" on the bank of a pond, lake, or river which commands an open view in all directions. There, in the decaying vegetable matter, under a spruce or a hemlock thicket, they will scratch out a hollow and curl up in a ball for a few minutes, knowing they can slip back into the water with one silent movement.

Before the ice comes, they sometimes pick a secluded spot on a marshy deadwater, tear up sedge grass with their paws, pile it into a stack until it forms a mound, then curl up on top for an otter nap. Other times, they catch a few winks in a bank beaver's burrow, or in any natural den with an underwater entrance. In winter, otter often curl up in an abandoned beaver lodge. I am convinced they even like to visit active ones. I suspect it's a one-sided pleasure. They must get the kind of reception a fundamentalist preacher would give a hell-raising, unsaved cousin.

To those naturalists, trappers, and nature writers who claim the beaver's greatest natural enemy is the otter, I say, "Show me." I am sure no man living has examined more otter scat than I have, nor analyzed it more knowingly. Occasionally, I have found muskrat fur, never that of a beaver. From what I have observed in a quarter-century, these two diverse water animals respect each other completely. Neither is a meat eater. Otter enjoy shore dinners; beaver are vegetarians. What is there to fight about? Only humans fight over ideas. If an otter were starving, of course it would attack and eat a beaver; but otter have an endless food supply. When would a civilized man eat another man?

Too often, we humans tend to assume that all wild creatures spend their lives hating, coveting, and killing as we do. Nature is neither vicious nor sentimental; she is honest. Whenever I hear a parent or a teacher denounce behavior by comparing it to that "of an animal," I sigh. If only we did behave like animals!

Because of its shape, an otter's head appears small, sloping back from the black nose to the diminutive, streamlined ears, set well to the side of the skull. Its coat-button nose sticks out beyond the underslung, sharklike mouth. Like the shark, it must rely on speed for its food. The large nostrils are situated on either side in such a way that, when "making knots" straight ahead, frontal pressure cannot build up. Its eyes are small, almond-shaped, and set far apart. I am convinced that, like a beaver, it has transparent eyelids. Many of the otter I have examined had died with their eyes open. Clearly visible was a half-closed, transparent eyelid, stretching across the eyeball. The enchanting whiskers are long, coarse, and silver gray.

The otter's short, stout legs also taper from the body to the webbed toes. Because of their unusual shape and strength, it takes a powerful trap to hold them. Unlike those of a beaver, the front, as well as the hind feet, are used for swimming, always keeping the forelegs well aft of the head and neck. Because its feet are on the side of the body, not underneath it, it swims with all four of them, aided by a powerful tail, in the same way a crocodile does. When the otter comes ashore, its neck reaches shallow water first.

The otter has strong, sharp teeth embedded in powerful jaws. Its four canine teeth are like a dog's, only much sharper; its molars so strong that, across seventy-five feet of water, I have heard them grinding up the bones and gills of coarse fish. They cannot eat underwater. Although I have never seen an otter in the wild hold a fish in its front paws to eat, I am sure they do. However, I have often watched them eat fish that lay on ice.

Like the beaver, an otter has two growths of fur. Its underfur is short, dense, and waterproof. The equally thick but much darker outer fur is longer and stiffer, but rarely more than one inch in length, being longest on the topside of the widest part of the tail; it is barely a quarter-inch long around the head, mouth, and tip of the tail.

The coat is much shorter than either a beaver's or a muskrat's. The otter needs to swim faster, and bulky fur would only slow it down. In the wild, its fur appears to be black, but it is really a dark, oily brown. I have seen only one coat that was a true brown without the usual black cast. The fur across the back is darker than that on the belly. No matter how dark the rest of an otter may be, the fur under its jaw and on its throat is a lustrous silver. The fur is darkest in November and December, when it has a beautiful, rich sheen. By the end of winter, after many weeks of sliding over the snow and ice during long daylight hours, the guard hairs appear silvery and develop a slight curl. In the fur trade, this is called "singeing." The pelt is no longer as valuable.

An otter hide is the thickest and strongest of any furbearer in the East. It is as thick as that of a deer, but tougher. That toughness, combined with its long, slender shape, induced the American Indian to fashion their quivers from otter skins. From all indications, an otter's lifespan is similar to a dog's, averaging about twelve years. Their muzzles, also, turn gray with age.

For such a large animal, its eyes are surprisingly small. Even more surprising is their keenness of vision. Gavin Maxwell, the English author who has written several delightful books about his domesticated otter, claims they have poor eyesight. This troubles me because it conflicts with personal conclusions based on my almost life-long experience with them in the wild. In situations where neither sound nor smell could have saved them, it seems to me sharp eyes did. They can see as well in dark water as they seem to be able to in bright sunshine.

Wild otter live primarily on fish, and these fish are not motionless. In southern New Hampshire and northern Massachusetts, they dart about in some waters the color of Coca-Cola, so dark one can hardly see one's hand submerged a foot and a half on a sunny summer's day. In the dead of winter, two feet of ice topped with three feet of snow covers the murky water. On the brightest day, it is midnight down there. Yet otter catch fish at will. I think they depend mostly on their sharp eyes, although it may be possible that this remarkable mammal is blessed with equipment similar to that of a bat, a porpoise, or a whale. Because they are so shy and their habits so secretive, nobody has experimented with the animals to find out. But it would not surprise me if a generous Creator has given them a fourth dimension. In any case, they certainly do have some way by which to locate small, fast-moving things in the dark.

One fall another link was added to my lengthening chain of evidence. Sign told me a pair would be passing through my territory about every seven days. Darkness was falling. Working against time, I set a trap more hastily than usual. Bending down so that my nose touched the water, I made a last-minute check to make sure everything looked natural. Satisfied, I left, confident it would hold an otter on my next visit. A week later, I approached the set from the opposite side of the deep, twelve-foot-wide stream. The clog, to which the trap was fastened, was still in place. Had my quarry changed their schedule?

Each week, until the ice came, I rechecked the set. I couldn't understand why the pair hadn't passed through. Curiosity drove me upriver a half-dozen miles, where their fresh sign told me they had indeed come through on schedule. Why hadn't they stopped?

Since an inch of ice now covered the river, I returned to pick up the bypassed trap. Even more important, I had to find out why the otter had avoided it. Something was wrong. Kneeling on the transparent ice, I looked down at the set. The trap was hidden, but off to one side something glimmered. I punched a hole in the ice with an axe, rolled up my sleeve, reached down and scooped up a loaded shotgun shell.

It had been almost dark when I set the trap. I wore a sleeveless canvas vest over a flannel shirt. The front pockets of the vest held shotgun shells. One had slipped, unnoticed, into the water, to be embedded upright in the mud. When the otter had approached, their small, sharp eyes had spotted a suspicious object, even though less than a quarter-inch of brass had been exposed!

Undoubtedly, Mr. Maxwell is right as far as his domesticated otter are concerned. Why should they depend on their eyes? They don't need to catch fish or to fear man. Remember, he observed captive otter in his home. I observed free otter in theirs. It is the only animal I know whose sight, hearing, and smell are equally acute. Other animals may have one, occasionally two, but rarely all three such highly developed senses.

While trapping mink as a boy, I often saw otter tracks in snow. Gradually, it dawned on me that these tracks appeared only at regular intervals, sometimes every seven or fourteen days, sometimes only once a month. Otter keep moving. They don't stay anywhere long enough to warrant a home address. That's one of the reasons why even people who ply a backwoods river or lake almost daily seldom see them. When the pups are born, these nomads do settle down, but only until their young are big enough both in size and understanding to travel along with mother and dad.

Far up the tributaries of the Ware River, I found where a pair of otter had come up to visit a backwoods pond, fished it for an hour, and left to go back down toward the main stream. The day was young. The tracks were fresh. Why not try to get ahead of the pair and set a trap for them?

I began to follow their trail. Fifteen snowshoe miles down the river I gave up. During the day I had passed several old dams where the otter had emerged to climb around before re-entering the river. The next morning, still intent on heading them off, I snowshoed directly to an old broken damsite five or six miles below the last one I had visited the day before. My quarry had already crossed over it. I stuck with them for another fifteen miles. Wherever spring holes bubbled along the river's edge, the pair had pushed up through the thin ice, surfaced, left their scat, popped back down the hole, and kept heading downstream. In two days, they had traveled more than thirty miles and were still going strong. They knew exactly where they were going. At that time, I could only guess.

Because the otter's pattern of travel can best be explained by using a particular river as an example, let's take an otter's tour of the Contoocook. This river is born quietly where Mountain Brook and the outlet from Contoocook Lake meet in Jaffrey. Finally, burgeoned by countless other streams both large and small, it flows north to be swallowed, full-grown, by the Merrimac, seventy-five crooked miles away. Ponds made by God, man, and beaver interrupt both the major and minor streams at irregular intervals.

A pair of otter will rarely explore just the main stream of such a river. If they do, it is an exception rather than the rule. Because they are happiest in primitive waters, they habitually leave the main artery to follow a tributary to its smallest capillaries. Unafraid there, they fish and play until both adventure and appetite are fulfilled. Returning to the main river, they swim on to the next branch, whether going up or downstream, to follow it to its beginning many miles and many minnows away. Thus a seventy-five-mile canoe trip on the Contoocook adds up to ten times that when traveled by an otter.

Otter know exactly where and how to travel, whether by land or by water. Many years ago, while hunting in compass country before any snow had fallen, I came upon some otter scat. This didn't seem logical where there was no pond, no river, no brook; only dense, dark forest. It so piqued my curiosity that, once I was back home, I paused only to take off my mittens before getting out my U.S. geodetic map. The nearest water was a small pond about a mile to the east, fed by a tiny brook that petered out many miles southeast from where I had found the scat. But when my exploring finger moved westward across the map barely a half-mile from my discovery it found another brook flowing northwest. This one emptied into a small, boggy pond. Here, in a dark swamp, another watershed was born. Somehow the otter knew it.

These navigators have demonstrated many times this ability to navigate, overland, from one watershed to another. The compass they carry in their shirt pockets is better than any I ever owned. When an otter journeys up the Contoocook, Ashuelot, Ware, Millers, or Connecticut rivers on a hundred-mile tour, he visits each of the dozens of tributaries that flow in turn from a hundred bogs and beaver ponds. When deciding to leave one watershed to go to another, he knows which brook to follow. There, with as definite a purpose, he crosses overland to the tiny beginnings of another river basin. What amazes me is, that in checking geodetic maps, I find the otter's overland routes, whether they be first- or many-time visitors, are, without exception, the shortest distance between two watersheds.

Because I kept an accurate census of the otter population on each of the watersheds within a fifty-mile radius of my home, I was aware of any new arrivals. They fished the entire system with typical zest. When the time came to leave overland, they always took the exact same route used by countless generations of others before them, even though no observable sign, no trail, marked the way.

Within seventy-five miles of my cabin I can take you to more than a dozen of these shortcuts connecting two major watersheds. Sometimes otter will use direct overland routes within a river basin itself in order to reach different branches of the same stream. Again, according to geodetic maps, it is always the shortest possible way. Why do they leave? How do they know where to go? Is this any less of a navigational miracle than that of the U.S.N.S. *Nautilus* probing its way under the polar ice cap? Biologists call it natural adjustment. Scientists call it instinct. I call it a gift of God.

Otter do not travel haphazardly. They plan their trips; and whichever one of their many tours they may choose to take, the itinerary is definite, never decided on the spur of the moment. That is why I can tell exactly where they'll be coming by and whether it will be two days or two weeks from the time the sign left on their latest stopover is studied.

Oftentimes, on arriving at a confluence, a pair of otter may split up, or one may leave the family group. This temporary parting is prompted by a gustatory rather than a marital difference of opinion. One, leaving its mate in the main river, ascends a contributory branch, passes straight through several fish-filled ponds, to arrive purposefully at some backwoods pool. It may spend two hours or two days in and about that tiny pond before returning downstream. Meanwhile, its mate has been a patient spouse, fishing the main river, but never too far from where they separated. Perhaps that parting conversation went something like this.

"Say, Sweetie, I got a craving for freshwater lobster and I'm heading for Lobster Pond. I know you don't like 'em, so you don't need to come along. Just stick around; I'll be back."

During their endless traveling otter may pass through a particular section on a certain river once a week, once a month, or once in a lifetime. There are a hundred places where they can come ashore. I call the coming-out places stations. For me, these otter stations read like a book. The scat explains everything: when, how many, how often, how big, what sex, what schedule, and whether the last restaurant served pickerel, perch, or sucker. Although otter leave clues for me, I never leave any for them. Nothing at the station should be changed. Above all, never use the rest room. One glance, one sniff, tells otter whether the enemy has been around. They'll never come back. Remember the shotgun shell!

The fall I was fourteen, I found some scat along a river. Thinking it the only otter sign in the world, I went to my old friend, Arthur Leonard, with my boy-shaking discovery. His quiet blue eyes studied me.

"I remember where that is, John."

Any otter stations, painstakingly discovered by me during the next decade, were no secret to this close-mouthed man of eighty. He never told me anything unless I had already discovered it, but he knew them all. He had trapped them himself only a few years after the Civil War, jolting from brook to brook, river to river, in his horse and buggy, with only his thoughts to accompany him.

At eighty, he was still lean and wiry. But his greatest strength was his obvious happiness with himself. It made him glow with a dignity and a kindness possessed only by people who truly like themselves. He was a silent man, in the way people who spend much of their time in the woods are silent. Chatting is a by-product of civilization. When he spoke to a boy bursting with unasked questions, his voice was gentle. The ruddy color of his weather-worn cheeks made a full beard look even whiter. His deft, though knotted fingers showed me how to skin my first otter. At eighty-eight, shortly before he died, he gave me the otter board he had fashioned himself in 1895 out of a special piece of basswood. On it was carefully inscribed data about the seventeen otter he had trapped during his lifetime. Now, both sides of the time-darkened board are covered with writing, for on it are recorded most of the otter I have caught over a span of almost forty years. If my cabin were burning, that board is one of two things I would risk my life for.

Whenever other people, including game biologists, write about otter, they emphasize the animal's affinity for sliding on chutes they have fashioned out of mud and clay along river and lake banks, where they "slide down, repeating the process over and over for the apparent enjoyment of it." Where? If only someone would take me to just one otter slide! After thousands of hours of research, I have yet to locate a single one in either north central Massachusetts, southern New Hampshire, southern Vermont, or along the entire wild length of the Allagash River in Maine, including the upper reaches of the West Branch of the Penobscot.

The closest to it was a clay bank which rose at a forty-five-degree angle, six feet above a deep pool in the Millers River. The otter slid down the slope, but only because it was the easiest way to get back into the water. There was no evidence that they used it as a playground. The natural places they choose whenever leaving a bank set steeply above water, show no signs of repeated activity, such as can be seen on any beaver "ramp," the path leading from a pond to the main "logging road," which these lumberjacks build on every log job. They use it to haul down their pulp. A ceaseless flow of traffic spatters layers of mud on either side of the highway.

Why is it that a self-taught naturalist like myself, who has spent better than 20,000 hours studying the river otter in its habitat, never saw even one otter "slide"? Yet almost every article written about this charming mammal centers about this "fact." What should I believe? What my eyes have read, or what my eyes have seen?

Under certain conditions, otter practice a peculiar, puzzling ritual. Arriving at one of their infrequent stations, they use their front paws to scratch up leaves, pine needles, and debris into a little mound from three to eight inches high. They deposit their scat on top. These mounds stand intact for many months before the elements can flatten them. Otter build these duff monuments only while on tour, never during the two to three months' interval when they are raising their young. Single otter always perform the rite. Because they are so friendly with one another, I believe it is one way they communicate.

Scat is not the only thing they leave on these memorials. At the base of its tail, underneath the hide, an otter has two glands, each about one inch in diameter. These sacs contain a viscid, yellowish-white fluid with a potent, tantalizing odor. The oil is ejected through two small vents near the anal opening. The secretion is not offensive like that of the mink, weasel, or skunk. But it is so penetrating that I had to be very careful not to puncture the glands while skinning the animal. If some of the liquid got onto my hands, the odor remained for days no matter how often or how hard they were washed. I am sure that, whenever an otter makes a mound, he ejects some of this oil on top of it.

No dog can resist the aroma. Even a mongrel can smell it buried under six inches of snow. He will dig his nose into it, flop over onto his back and, with a silly look on his face, roll wildly around in it until he is dragged away.

For a long time, I have pondered the otter's reasons for building these mounds. I believe the same emotion impels the otter which makes a climber build a rock cairn on a lonely mountain. Perhaps a pair is saying, "We have been here. The fishing's not what it used to be." Or a single one pleads, "I am lonely. Come join me."

15. The Dog Otter
Is a Family Man

The dog otter is a family man. He chooses one mate, travels with her year round, and when she bears his young, shows strong fatherly consideration for them. He assumes his full share of the family load, for the otter's responsibility for its young has no gender. The dog enjoys his pups and teaches them with firm kindness. Here is an extraordinary family relationship. Otter were practicing togetherness long before marriage counselors started preaching it.

In the Northeast, all other adult fur-bearing creatures except the beaver live together only during the few frantic days of the rut. Whether it be July, September, January, or March, I have always found otter living together as a family group. They travel together, they fish together, they eat together, they play together, they grieve together.

Beaver are also family creatures but for different reasons. One beaver can't build a dam any more than a single army engineer can. Who ever heard of a one-man construction job? Beaver live together through necessity; otter, by choice.

Otter are happy animals who enjoy living. They are fun-loving but not frivolous, frolicsome but not foolish, irrepressible but not irresponsible. The inference that they are "good-time Charlies" is unfair. Like any emotionally mature adult, they bear their burdens cheerfully.

Even during the rut, otter do not fight among themselves, and a marital triangle is unheard of. Even though theirs is a "live and let live" policy, if challenged, their courage is unlimited. No single domesticated dog would be a match for those savage teeth, that mongooselike agility, and sheer guts.

I believe the otter's rut begins in the middle of February and ends about the middle of March. This deduction is based on the sudden, drastic change in their behavior during those weeks. Only nature's most powerful urge can change shy, cautious creatures into bold, impulsive ones.

Some accepted authorities claim the female otter has a variable gestation period of from 280 to 380 days after mating "in the spring." This would infer that the young would be born sometime between the following December and April. This puzzles me. Based on a lifetime's observation, the female otter definitely does not bear young every year. Of the hundreds of otter I have observed, I have yet to see a family with siblings of different ages; but yearlings traveling with mother and dad are usual. Furthermore, based on the size of the smallest pups I have examined during the months of November and December, I would say they had more likely been born in May or June.

Over a period of thirty-five years, during the months of November and December, I have examined dozens of female otter, ranging in age from yearlings to matriarchs of eight or nine years. None of them showed any signs of pregnancy. The overwhelming majority of them were married.

Mrs. Otter generally gives birth to her young in an abandoned beaver lodge. Sometimes she may pick a den in a riverbank, but the maternity ward must always have an underwater entrance. Above all, the female chooses her delivery room as far away as "animally" possible. She usually gives birth to two pups. During four decades of field work, I have observed only a half-dozen triplets. The young remain with their parents for approximately a year and a half before venturing out on their own; there is much to learn. Courses on "How to Avoid Highways," "How to Steer Clear of Duck Hunters," and "The Identification of Enemy Traps" are stressed.

During the mating season, single or widowed otter become obsessed. My first experience with one in search of a mate amazed me. In the early fall of 1928, the dam went out in a small, then primitive pond, which was the headwaters of a large river basin. One morning, the following November, while trapping mink, I heard a commotion in a deep pool below the broken dam. An investigation revealed fresh otter sign near a den under an overhanging bank where a pair had set up temporary housekeeping. The fishing was too good to leave. The one otter trap I had with me was set. The next day it held a big female.

During the following January and February, while checking mink traps, I snowshoed up and down this same brook. There was no otter sign until early March. One morning, shortly after daybreak, I approached the brook a half-mile below the broken dam. Within hours, a large single otter had gone upstream and come back down. Intrigued because both sets of tracks had been made less than an hour apart, I followed the trail. The otter had headed straight for the den under the bank. It had checked it thoroughly. Then, without hesitating, it had headed back downstream. During the next two days, I followed its trail down the main river for more than thirty miles. It did not fish. It did not play. It did not visit any of a dozen tributaries. It had come back up the river with but one question in its mind. It had found the answer.

If land and freshwater mammals held a swimming meet, every medal, gold, silver, and bronze, would be copped by otter. The beaver also swims to live, but where an otter is swift and powerful, the beaver is only powerful. The one is a tugboat, its wide, sturdy hull powered by two oversize paddles, pushing and pulling heavy timber about its do-it-yourself canals and ponds. The other is a racy destroyer, propelled by four screws and, aided and guided by a streamlined rudder, constantly harassing and intercepting convoys of fish.

These speedsters can swim underwater for almost a half-mile. Of all the freshwater mammals, they can stay underwater the longest, and a surprised otter can make a getaway worthy of a commando.

Rain had swelled the river at Barre Falls before a sudden cold snap froze the sides of the stream as well as the flooded swamps bordering it, leaving twenty yards of open water in the middle. Crashing up along the shell ice, I heard something else smashing through in the dead grass ahead, followed by a loud splash. Running to the river, I saw what had happened. While getting breakfast, an otter had also climbed up onto the shell ice covering the bank. Hearing me, he had bounded back into the stream. In the few seconds it took me to reach the river, his head reappeared 900 feet downstream. Confident again in his own element, he reared up with half his body facing me. Wary as he was, he couldn't resist looking back.

Otter find weak places in ice, lunge up, break through with their heads, and emerge. How do they identify these weak spots?

To me, they look no different from the stronger ice. Obviously, their know-how is built-in. When a pond or a lake is solidly frozen over with several feet of ice, they resort to a subway. Erosion forms tunnels around large tree roots that extend into water. Otter know every one of them. They twist and turn upward along these subterranean, water-filled channels until they suddenly pop out of a hole in the ground, sometimes a dozen or more feet from the pond. Many times, I have followed the subsequent trail along the shore for a considerable distance before it again suddenly disappeared into the ground. They seem to know exactly where to find the entrance to a tunnel leading into a pond, even though it may be covered with several feet of snow.

Only during the rut are they likely to traverse the full lengths of backwoods ponds and lakes, stopping at islands and on peninsulas. Even when compelled by nature's most powerful urge, they still prefer to travel during a storm when the ceiling is zero.

In March, 1967, a large single otter journeyed up one such deserted pond for two miles. It left its tell-tale trail like a causeway on the windswept snow, before vanishing into a hole under a clump of buttonbush on a tiny island. The neat opening in the snow seemed too small to accommodate the otter's solid body, but if the whiskers go through, so will the rest.

A day after this sighting, twenty miles away on a different watershed, another bachelor came out below a small beaver dam beside a secondary road, left his calling card, then traveled overland for four hundred feet. Three-quarters of the way up, he decided to cross the highway, but at the very edge of the blacktop, he changed his mind. He skied another hundred feet along a snowbank before turning around and going back to the beaver pond. He gave me the definite impression that he had never been there before. He entered the pond above the dam via a hole barely eight inches across at its widest. Because the length of impressions from the tip of the forefeet to the tip mark of his tail averaged forty-five inches, this dog was of marrying age. The average length between jumps was twenty-five inches. While traveling away from the pond, his jumps were a uniform, hands-in-the-pockets saunter, but, on the return trip, something startled him and the jumps became erratic. For over fifty feet the familiar dot-dash-dot imprints of an otter in a hurry marked the corn snow. In all, he made five slides measuring from four to sixteen feet, the longest being a traverse of a sudden declivity demanding the skill of an Olympic bobsled driver.

Awkward on land, otter leave a strange trail in mud or in clay, with a series of paw imprints, but with no body mark, as if a giant inchworm had measured its way along the water's edge. Evidently, they don't like to get mud on their pants.

Having been designed for water travel, the short legs are not meant for snowshoeing. However, these prodigies know how and when to travel overland in winter. Whenever snow is deep, hard-packed, and topped by one to three inches of powder, an otter races along by employing gymnastics, shifting into overdrive, and having fun doing it. It accomplishes this by taking two or three energetic, power-building jumps, each about two or three feet long, depending on the size of the animal, then diving head-long onto the snow with stubby front legs tucked underneath, and body, hind legs, and tail splayed out behind; exactly like a child who picks up a sled and runs with it before flopping it down to get a faster, longer ride. Thus otter slide forward effortlessly, until the momentum slackens, whereupon they leap to their feet to repeat the jumps and slides. The length of each slide varies with the snow conditions. If the snow is perfect, so's the sliding, for nature has equipped them with built-in "flying saucers."

While striding along the undeveloped shores of Stony Pond, I noticed an otter's ski tracks a few yards ahead. Leaving one of the inlets, it had skidded the entire 1½-mile length of the pond. This rare opportunity to measure the length of a slide delighted me. My size ten boot, toe to heel, went twenty-one times along one slide. That's quite a broad jump, good enough for competition! No more than two preparatory leaps linked each slide to the next.

Like any skier, otter take advantage of gravity by schussing every slope going their way. At the top of a hill, they throw themselves forward. Since their steering wheels are designed for water rather than for snow travel, as they gather momentum, they often whirl right into a tree, ricochet off, and continue down to the foot of the hill. Because they are so supple, tough-hided, and well muscled, they don't get hurt. I have often measured downhill schuss marks 75 to 100 feet long. Their paternalistic pattern of family living also carries over onto the ski slope, for when cross-country skiing, the family goes in single file, with the dog breaking trail for his wife and children.

The hungrier an otter is, the faster it swims. One morning, as my canoe slid silently around a bend of a river, ripples wrinkled the smooth surface ahead, signifying the presence of a swimming animal. An otter's head broke the surface, then another's. I decided to watch until they saw me. The stream of bubbles, constantly coming to the surface, gave me a tracking pattern. They appeared, first in a straight line, then veered suddenly to either an abrupt left or right, ending only when a wet skull popped up with a fish crosswise in its triumphant mouth. That day on the big deadwater proved something I had suspected for years. An otter can outswim any fish, anytime, anywhere.

Otter eat whatever variety they find. In southern New Hampshire and Massachusetts, that limits them to suckers, dace, perch, sunfish, minnows, pickerel, and pout. Fish bones and fish scales predominate in all their scat. A quick glance tells which piscatorial delight this native Isaac Walton has been eating. Are there the vulgar scales of the sucker, the shiny ones of the dace, the black pulp of the hornpout, the amber scales of the trout, or the tough, pink shells of the crayfish? Ninety-nine per cent of otter scat I have examined during the last four decades contained the scales of coarse, warm-water fish—very rarely the delicate scales of trout. This is not a matter of taste. Otter like trout as much as

I do. But, like me, they have to settle for what there is. Because our brooks are slow and sluggish and our ponds shallow, the waters are warm. Trout do not live in warm water. Otter do not fish brooks near roads, even though they may have been stocked with trout. Highways and otter do not mix. If one chooses to travel up one of these streams, it does not stop to fish. It only uses the waterway to reach a secluded pond where it will be safe from man.

Most fishermen seem to think otter menace our trout population, citing the scarcity in ponds reclaimed by the state; for anglers rarely catch a trout after the first of May. "The otter get 'em all!" they argue.

Have you ever fished a reclaimed pond on opening day? If you arrive after six in the morning, you'll have to park your vehicle a half-mile from the public landing. Even then, an hour later, you'll have your limit. Everyone does. On the second day, you'll catch trout, hooked the day before, whose jaws are too sore to permit a second escape. Don't go back the third day. The trout are all gone—until next year. This is strictly a put-and-take proposition: "we put them in today; you take them out tomorrow," with the stocking taking place as close to opening day as the ice permits, so that the fish will survive. It would be sheer coincidence if an otter, on his regular route, arrived at a reclaimed pond in the few hours between the departure of a hatchery truck and the arrival of the rubber-shod hordes.

Even if otter knew when the trucks were coming, they would not be there to welcome them. Most of these ponds are completely surrounded by cottages. A scant fifteen years ago I could have taken you to a dozen marshy ponds and showed you fresh otter sign. Today, those mudholes have been reclaimed by land developers, with row upon row of camps. An otter hasn't been there since the first nail was driven.

Even if an otter were to get into a fish hatchery, it wouldn't stay. The taste of trout is not worth the sight and smell of men. If these fish lovers live in natural trout country, of course they will eat them. Why not? Aren't they the natives and we the outsiders? But because they fear man, they avoid his roads and his dwellings. Trout and otter share a common enemy.

Like any good Yankee, the otter makes do with what there is. Paradoxically, he can also be a gourmet, often traveling miles to reach a certain pond offering a favorite tidbit: a painted turtle or a crayfish, the freshwater lobster already alluded to. He relishes frogs, and, on rare occasions, will eat a muskrat. Inexperienced zoologists often think the fur they sometimes observe in otter scat belonged to a beaver. It is muskrat. Even though the entire animal is eaten, the fur passes through the intestinal tract intact. Obviously, some sort of natural therapy is involved; the fur scrapes off fish scales stuck to the plumbing.

Once, when a light dusting of snow had transformed the clear river ice into alabaster, I was walking gingerly down a frozen river. As I came around a bend, dark splotches discolored the snow-covered ice ahead. A pair of otter, passing through the night before on their regular itinerary, had slithered onto the ice through a hole formed by the warm flow of an underwater spring close to shore. The water temperature here was above freezing. Even when surrounded by thick ice, spring holes remain open throughout the winter, enlarging during a thaw and becoming smaller whenever extreme cold follows a heavy snowfall. An otter's built-in compass points magnetic north to every such spa.

These two had dug up a hibernating turtle, to eat only those parts protruding beyond the shell. Hoping that terrapin on the whole shell might spice their midwinter menu, they had rummaged around in the mud at the bottom of the spring until one of them had latched onto a snapping turtle. Clambering out onto the ice to relish the tidbit, they shook themselves, splashing mud onto the snow. Otter have a weakness for their own kind of hors d'oeuvres, and never bypass a potential snack bar.

I have examined the remains of many such turtles, most of them eight to ten inches in diameter. This one barely measured three inches across. After bringing it up onto the ice, the gourmets had decided it wasn't worth a cracked molar and left it there. The wee turtle crept ten halting inches before freezing to death, its tiny head, tail, and diminutive feet extended. I still have the turtle shell, with an otter tooth mark perfectly preserved. It's the only time I ever saw turtle tracks in the snow.

Every time I fish white water, I remember a pair of otter who taught me something. Although it was the end of November, no snow had fallen, and the section of river consisted of a series of frozen deadwaters 75 to 100 yards long, connected by short stretches of white water. Approaching the river, I discovered the still steaming scat of two otter. Hoping to catch a glimpse of the pair as they headed upstream, I loped through the woods for a couple of hundred yards before cutting back to the water's edge. No luck. After three such sallies I finally came out just below some rapids. Hearing a commotion a short distance upstream, I swung about to look.

Two dark bodies thrashed and boiled in the white water. One burst to the surface with a twelve-inch sucker in its mouth. It rode the rapids downstream, bobbing with the current until reaching the frozen deadwater, where it slid swiftly and smoothly onto the ice. Almost at once the second otter followed, a sucker jutting out beyond its whiskers. They ate dinner on the icy table. With the last crunch each slid into the quiet pool below the rapids. Alarmed, and seeking safety, the suckers raced for the white water upstream, the otter hot on their tails. What a chase it was, beginning with a hit-or-miss, upside-down, inside-out pursuit, always ending with a fish crosswise in an otter's mouth. The otter knew that in fleeing the white water, the fish would bunch up, making them easier to catch. With each personal encounter, my respect for nature's Ike Walton increased.

Here is a mammal which takes seriously the old adage about "all work and no play." Because I knew it still remained one of their favorite retreats, I approached the spongy northeast shore stealthily. Among the stumps that pincushioned the boggy pond, a family of three otter were noisily champing down a pout supper. I don't know how long they had been fishing, but as I arrived they suddenly started to chase one another over and under water-soaked logs and decaying dead trees floating around the stumps. The players of this follow-the-leader touch football got going so fast they flowed over the bobbing obstructions like so much animated syrup. When the one being chased turned abruptly to egg on his pursuer, I swear he was grinning. Then, just as suddenly, they changed roles. Surfacing like a seal or a porpoise, this Yankee cousin catapults skyward high enough to show most of its sleek belly, with the chunky forelegs neatly folded under. Watching until the otter family disappeared in a cove, still rollicking, I knew none of their clan could ever develop ulcers or hypertension. Ever since that autumn day, whenever feelings of guilt about finishing a stone wall or cleaning the cellar assail me, I remember those happy acrobats and reach for my fishpole or my snowshoes.

Because they are intelligent and sensitive, otter are also wary. Having learned in their youth that man is their only enemy, they remain an animal of the backcountry. Whenever a village with its inevitable dam interrupts an otter's journey up or down a river, he hesitates. Sensing danger wherever buildings stand guard on either side of a dam, he stops a half-mile above the barrier to wait for a stormy night. Only then does he leave the river and head for the woods, bypassing the settlement, to re-enter the protecting cover of water a quarter-mile beyond the last building.

In primitive backwoods, these doubting Thomases will travel overland from one watershed to another in the daytime, but only if the sight and scent of man is not there. Even when the origin of another river basin can be reached by crossing a little-used country road, if human dwellings watch nearby, an otter hesitates.

One stormy March day, Jiggs and I were fighting the north wind back to our truck after a fruitless day. No one else had visited McKittrick Mountain since deer season. Four feet of packed snow carumped beneath my snowshoes.

Coming back on my own webbed trail, I was following a tote road that balanced on a razorback ridge so neatly I could look down through the woods on either slope. Off to one side I noticed a disturbance in the snow. Three or four hours after I had passed by that morning, a family of three otter, lunging uphill through the deep snow toward the razorback, had approached within fifty yards of my trail. As soon as my odor was detected, they bounced back along their own tracks. Then, paralleling the road for a hundred feet, they tried again to cross. My scent still barred the way. Turning tail, they slid back over their tracks once more. Then, for the third and last time, they tried to find an unguarded crossing. My phantom still hung in the air. Not once had they come close enough to see the snowshoe trail.

They tobogganed back to Hedgehog Pond, a mile distant, to wait another storm. I have known otter to bide their time for as long as seven days, awaiting the right traveling conditions before leaving one watershed to reach the beginnings of another.

Many naturalists allege that otter and beaver do not get along, contending that the former will even kill and eat their hard-working relatives. I do not believe it. The proof can be found in beaver ponds and in otter scat.

Before the recent reintroduction of beaver into southern New Hampshire, many small tributaries fed rivers, ponds, and lakes. Although these streams were full of fish, otter never visited them. They didn't dare. How could a large animal hide in shallow water?

When their engineering cousins surged up those brooks, intent on building dams, a New Deal began for otter. Soon, series of dams, each holding back a pond, were strung along the waterways, like a chain of fish markets. The otter patronize every one.

They also enjoy their cousin's lodges, popping in occasionally without warning for a brief nap. With their "twinkle-in-the-eye" attitude toward life, otter cannot resist pestering their "stick-in-the-mud" benefactors. Because they can swim circles around the single-minded drudges, there are plenty of chances between the poplars on shore and the feedpile beside the lodge.

Ironically enough, it is the current beaver trapping laws that will eventually eradicate the otter in New Hampshire. "And what has beaver trapping got to do with eradicating the otter?" you ask.

In comparatively recent years there were no beaver in the southern counties. The beaver the white settlers found were all trapped out. However, many well-established colonies flourished in our nothernmost counties, under the strict supervision of the Fish and Game Department. In 1940 the Department began live-trapping beaver in pairs and transplanting them to the southern counties, with the same protective laws applying to the emigrants. Being true pioneers, the settlers adapted quickly to their new world.

By 1944 the original nucleus had multiplied so that the Fish and Game Department decided to open a short trapping season, while still maintaining deliberately restrictive laws. Part-time trappers dug out their rusty equipment and went out after the pelts. Those were the days when a blanket pelt sold for sixty-five dollars. They caught beaver, but because of the strict regulations, failed to thin out the dam builders. Meanwhile, these were becoming a nuisance in many sections. Their dams plugged up culverts; their backed-up ponds flooded cultivated meadows. A contributing factor to the problem, other than that of the beaver doing what comes naturally, was the sharp decline in the value of their pelts. Milady had discovered man-made "beaver" coats at seventy-five dollars apiece.

About this time, land damage complaints from angry land-owners and frustrated road agents flooded the Fish and Game Department, which felt compelled to relax the beaver trapping laws. Among other things, traps were allowed to be set on beaver dams themselves. Most trappers, whose experience had been limited to muskrat and mink in the open waters of November, knew little, if anything, about the otter. Lacking woods experience, they didn't even realize such an animal existed. During the initial short, midwinter seasons of January and February, the accidental taking of an otter in a beaver trap happened so rarely that it made no inroads on the meager otter population. But when trappers were allowed to set their traps in the spillways of the beaver ponds, they began catching an occasional otter.

Because the otter's mating season begins in the middle of February and extends into March, his defenses are down. This prudent gentleman of the waterways, with his keen eyes and superior senses, can easily avoid the traps of amateurs during November, December, and January. But, intent on mating, he falls easy victim to the same clumsy trapper during the rut. At that time, except for the spillways, beaver ponds are still frozen solid. The only way an otter can get into the pond is over the spillway. Usually quick-witted and vigilant, he doesn't differ much from man when it comes to thinking with his glands instead of with his brains.

Having a beaver by the tail, the desperate Fish and Game Department has extended the trapping season more and more until, in our southern counties, it is now legal to trap beaver for as long as four and a half months. And that isn't all. Wherever and whenever land damage complaints occur, the local conservation officer assigns some spare-time trapper to clean out the troublemakers, be it June or January. I know one such trapper who operates until June; he'd trap all summer if his neighbors weren't so stuffy about the skinned carcasses steaming in the summer sun.

For more than twenty years I have lived in ideal otter country. It is crisscrossed by hundreds of miles of rivers, brooks, ponds, and bogs. I am willing to bet a 1945 dollar that there are never more than eight fishing those waterways at any given time.

Whenever an otter loses its mate, it searches desperately for its spouse for weeks, even months. Then it begins resolutely to look for another. For the past four years, a large otter, who is working a watershed that covers fifty square miles, has been regularly visiting one of its favorite haunts within three miles of my cabin. It is alone. From its size and habits, it must be a male. Several times each winter, during a storm, he leaves the river to travel overland about a half-mile to a small pond. He does no fishing. He follows the west shore for a little way, then comes back. Sometimes, in the early spring, he disappears for as long as six weeks. I am sure he goes to another watershed, still searching for a mate. My alarm is justified. I have never before known an otter to live alone for more than a single year. Why has this one remained mateless for four? No otter is a hermit by choice.

The crux of the problem lies in the fact that beaver ponds and otter go together. As long as beaver build ponds, otter will fish them. As long as it is legal, on the one hand, to trap otter throughout their rutting season, and also legal, on the other hand, to set traps on the spillways of beaver ponds, the future of the otter in New Hampshire remains in jeopardy.

Neither the Fish and Game Department nor the Legislature understands what damage they have wrought and what inroads have been made on the already scant otter population. By extending the beaver season into the late spring and making the taking of otter pelts legal, the department, whose purpose is to conserve, is unwittingly destroying.

If it would do any good, I would gladly go to the State House, and on my knees, plead, "For God's sake, gentlemen, while there is still a little time left, save the river otter!"

16. Home Is
the Hunter

I was nine years old the first time I went hunting. It was a bright, late October day when, with a fifteen-year-old companion who owned a single-barrel twelve-gauge shotgun, I hiked five miles through leaf-strewn woods, beyond the outskirts of the town where we both lived. In his pocket, my comrade carried three shells. As we approached a tremendous oak tree, bristling with branches, and six feet through at the butt, partridges, startled by our appearance, began to whirr away from their roost. Seven birds had disappeared by the time my companion got his cumbrous weapon to his shoulder and pulled the trigger. The gun misfired. Four more grouse emerged. Cliff tried unsuccessfully to fire a half-dozen more times. He held out the shotgun. "You try it, John."

To me, the gun looked like the cannon in the town square. But, somehow, I hauled it up into firing position, closed my eyes, and pulled the trigger. Again, it misfired. I pulled the trigger a few more times before I began to count partridges. I do not know how many had flown while my eyes were squeezed shut, but twenty-seven more ruffed grouse, in addition to the first eleven, flew from the shelter of that huge oak.

After my naive introduction to the art of grouse hunting, it took close to four years before enough muskrat had been trapped and their skins stretched and sold to an itinerant fur dealer, to buy a future dyed-in-the-wool bird hunter his first shotgun. Still young and a poor wing shot, my enthusiasm didn't ebb; I could always bag my limit under old cider trees on the abandoned farms that dotted the local landscape. Invariably, my first wing shot would miss. But, too drunk on apple juice to fly high and straight, the inebriated grouse would crash into bushes, and while they were untangling their landing gear, a boy had time to reload.

Today, a housing development stands where that great white oak harbored a host of grouse. The town is now a city that sprawls for miles beyond what was once forest, field, and marsh. And, far north, I can hunt all day in ideal cover, and never flush a single grouse. I haven't seen a drunk partridge in years. I'd gladly settle for a shot at one grouse a week, a grouse that was stone-cold sober.

Just as the frontiersman tamed the West, so the gasoline engine has tamed the East. It took the former blood, sweat, and a hundred years. The latter did it in less than twenty. Oil plus money is high-powered fuel. Ten years ago, armed into burgeoning battalions of motorboats, camper-trailers, airplanes, four-wheel-drive vehicles, and snow machines, the gasoline engine drove me out of the woods. The vast private kingdom I had once called my own was overrun. In today's woods, a hunter on snowshoes is as out-of-date as a crossbow at a missile site. As gracefully as I could, I hung up my traps and my guns.

Until it met its mechanical Waterloo, my domain had been the area's last wilderness. For years, from the day after deer season ended, until the snow melted, I had run the remote ridges with a hound as my only companion. But we were never alone. Our realm had teemed with game: grouse, deer, mink, beaver, otter, and bobcats. Deer were everywhere. In fall and early winter, secluded river crossings looked like barnyards. A dozen winter yards, some harboring as many as twenty deer, were scattered throughout the area. Less than ten years ago, from a rocky pinnacle, through binoculars, I had watched thirty-two deer as, prodded by nature's urgent elbow, they began to leave their separate yards. Weak from hunger, and feeling the cold more quickly each night, they were following the sun toward the southern slopes of the mountain. There, sunshine waxing warmer every hour would soon expose the top shelves of nature's pantry.

Today, when I strap on my snowshoes and head for the familiar ridges, it's as though I were going to a wake. But I am drawn to the wooded mountains as though to a dying friend. As I climb, my best memories rise to greet me. But now I am alone. My woods are dead, overhunted, barren of game, with only snowmobile tracks where deer and bobcats once walked. There is often not a single deer track through eight miles of what many would still consider wilderness country. Huge choppings, an aftermath of recent logging operations, offer tender, juicy red maple shrubs. They stand unnibbled, while verdant ground hemlock remains untouched. Otter no longer explore the tributaries of a once isolated lake, or travel up to fish beaver ponds and to catch a few winks in beaver lodges. Few rutting mink race up and down the brooks and rivers. Wealth, progress, and high-octane gasoline have taken their toll.

Once remote lakes and ponds, where only deer, ducks, otter, and I trespassed, are now surrounded by layers of summer camps, made accessible in winter by snow machines. Points of land, where generations of otter have disembarked since primeval times, now support camps with wharfs jutting far out over the water. Two-lane roads dissect filled-in swamps that less than a decade ago knew only muskrat houses.

Lakes, once isolated by the north wind, are assaulted by a new breed of ice fishermen. One day, while crossing a thickly frozen, snow-covered cove in a lake that still harbors a handful of salmon and trout, I was dumbfounded to see a dozen holes with tackles, but with no men's tracks leading to them. Near shore, the fishermen were playing poker in a heated camper. Whenever a flag unfurled, they shifted their private club into low and drove across the ice from hole to hole to check their catch. It has been rumored that the automobile industry is designing a special camper for future ice fishermen, so that never again will they have to leave the shelter of their vehicles. Still in pajamas, they will be able to drive out onto ice, press a button, and a hydraulically operated auger will drill their holes.

With the ever burgeoning numbers of hunters in the woods, it is ironic that the greatest killer of them all, for whom there is open season on all species of wildlife anytime, anywhere, is the "owner" of most Americans, the automobile. In 1968 alone, according to a scientific tally taken on our national highways, over 1,200,000 wild creatures of every species died under the wheels of speeding cars. We can only speculate as to how many crawled away to die.

The social revolution that has disrupted our affluent society during the last fifteen years, has spilled over into the woods. But when I try to explain what has happened to a frustrated youth in his twenties who can't understand why there isn't more game for him to shoot at with his custom-made rifle, I feel like a tired old fox hound trying to describe the good old days to a skeptical French poodle.

For six years I have been the delighted owner of three hundred yards of a brook. Cold, clear, spring-fed, it has always hosted trout. Naturally, today, it needs supplementing by the state, and stocked trout are dumped into the stream from a narrow bridge near my cabin. I am extremely fond of "brookies," even when they taste of hatchery liver. Summer evenings, I dash home, grab my fly rod, and sneak down to cast into likely looking pools. It is useless to cast downstream; a black gnat can't penetrate the solid aluminum canopy of beer cans bobbing on the surface. But, upstream, where the current clears away the debris, I cast out hopefully. Perhaps I'll get a strike from a squaretail lurking inside a submerged rubber tire.

I take pride in my ability to hunt and to trap. I learned it long, hard, and well. I still prefer deer meat to beef, and partridge to chicken. I still revel in matching wits with the wild. The law tells me I may shoot one deer of either sex during a twenty-five-day season; that I may trap beaver, otter, and mink for five months of the year, and fisher for four; that I may continue to kill bobcats for a bounty. But, to me, to hunt is to pursue plentiful game in season. I do not want to shoot the last deer in the county, or to trap the last otter in the Contoocook watershed.

My privilege to hunt comes loaded with responsibility. I owe the game I pursue a place to live, food to find, and ample cover to hide from me. Even more, as a man, I owe them an even break. A rifle or a shotgun and my own two legs are the only accessories my self-respect will allow. An animal's built-in instincts, reflexes, and four-wheel drive balance the scales. For me to shoot from a car, a motorboat, an airplane, or a snowmobile would be to demean myself as a man.

Hunting can never again be what it used to be, but it can be better than it is. If only I could persuade the powers that be to shorten seasons that are too long, to close seasons whenever breeding stock is threatened, to use a fluctuating buck law, to have the courage to manage their department at eye level with the needs of wild creatures, with never a downward glance at political toes being trampled. We can look to the men who first hunted these ridges and valleys for our answer.

The American Indian was a true hunter-conservationist. For centuries, he cooperated with nature. He respected the creatures that fed and clothed him. In writing about the culture of the northeastern Indian hunter, J. M. Cooper tells us: "Their religious teachings permitted no wasteful or arbitrary destruction of life in any form. In taking animals for food and for skins, the survivors, as breeding resources, were constantly considered. . . . Ages of harmonious co-living with the life of the forest, swamps, and ravines of the Eastern woodland, left them the sense to accord to all forms of life the right to live, to propagate, and to fulfill their own destinies, as man himself claims it. This assertion is made with no degree of exaggeration. It is rather understated. . . ."

The American Indian hunted these hills and trapped these rivers countless generations before me. According to our cultural standards, he was a savage. Yet, centuries ago, he learned to join hands with Mother Nature, and, as far as animals were concerned, to think like God.

How can I do less?